AF572527

Nanotechnology

An Introduction to Synthesis, Properties and Applications of Nanomaterials

Thomas Varghese
K.M. Balakrishna

7/22, Ansari Road, Darya Ganj, New Delhi
Tel.: +91-11-4077 5252, 2327 3880
E-mail: orders@atlanticbooks.com
Web: www.atlanticbooks.com

Published by Atlantic Publishers & Distributors (P) Ltd., 2024
Reprint 2019, 2020, 2021, 2023, 2024

Disclaimer

- The author and the publisher have taken every effort to the maximum of their skill, expertise and knowledge to provide correct material in the book. Even then if some mistakes persist in the content of the book, the publisher does not take responsibility for the same. The publisher shall have no liability to any person or entity with respect to any loss or damage caused, or alleged to have been caused directly or indirectly, by the information contained in this book.
- The author has fully tried to follow the copyright law. However, if any work is found to be similar, it is unintentional and the same should not be used as defamatory or to file legal suit against the author.
- If the readers find any mistakes, we shall be grateful to them for pointing out those to us so that these can be corrected in the next edition.
- All disputes are subject to the jurisdiction of Delhi courts only.

Printed & bound in India by Atlantic Print Services

Preface

Nanotechnology is the science and engineering of making materials, functional structures and devices in nanometer scale. One nanometer is equal to 10^{-9} meter. In Greek, *nanos* means dwarf and *technologia* means systematic treatment of an art or craft. Nanostructured inorganic, organic, and biological materials may have existed in nature since the evolution of life started on Earth. Nanotechnology deals with materials or structures in nanometer scales, typically ranging from 1–100 nm. Size reduction can lead to a whole range of new physico-chemical properties and a wealth of potential applications. However, access to these nanostructured entities requires the development of suitable methods for their elaboration.

The concept of atomic precision was first suggested by Physics Nobel Laureate Richard P. Feynman in 1959, in his legendary speech at the California Institute of Technology. He stated: "The principles of physics, as far as I can see, do not speak against the possibility of maneuvering things atom by atom...." Nanotechnology has become a very active and vital research topic which is rapidly developing in industrial sectors and spreading to almost every field of science and engineering. Several major research and development programs on nanostructured materials and nanotechnology have been launched by governments worldwide. This field of research has become an area of great scientific and commercial interest because of its rapid expansion to academic institutes, laboratories and industries.

The research on nanotechnology is evolving and expanding very rapidly, and hence it is impossible to cover all the aspects of this field in one book. Thus, the aim of this book is to summarize briefly the fundamentals and established techniques of synthesis, processing, characterization, properties and potential applications of nanomaterials so as to provide important information on nanotechnology to the readers.

The book is divided into seven chapters. Chapter 1 is an introductory chapter. Chapter 2 gives an overview of various types of nanoscale materials, their properties and significant foreseeable applications. Chapter 3 explains how properties and behaviour of materials change as their characteristic dimensions are reduced. A concrete knowledge about the properties of nanoparticles with particle size is essential for the understanding of the fundamental concepts of condensed matter. The same has been provided in a systematic way. Besides, the concepts of scaling laws are also discussed in this chapter.

Synthesis and processing of nanomaterials and nanostructures are the essential aspects of nanotechnology. Studies on new physical properties and applications of nanomaterials are possible only when nanomaterials are made available with desired size, morphology, crystal and micro-structure, and chemical composition. In chapter 4, different techniques for the synthesis of nanomaterials are presented. Various structural characterization methods that are most widely used in characterizing nanomaterials and nanostructures are discussed in chapter 5. In chapter 6, various carbon based nanomaterials are discussed along with their properties, structure and potential applications. The last chapter intends to provide some examples to illustrate the vast range of potential applications of nanomaterials.

The book would serve as a general introduction to people just entering the field of nanotechnology, and also for experts seeking information on other subfields.

Thomas Varghese
K.M. Balakrishna

Acknowledgements

We are extremely grateful to Dr. Elby Titus of Aveiro University, Portugal, Prof. Lissi S. Plathottam, Dr. Shaju Thomas, Prof. Susan Varghese, Dr. Sanish P.B. and Dr. Girishkumar G.S. of Nirmala College, Muvattupuzha, Kerala, for their help, useful suggestions and discussions. We also wish to thank Ms. Susan Kurian and Suraj V. Thomas for the help and support which they provided during the writing of the manuscript.

We wish to put on record our sincere thanks to the Principal, staff and management of Nirmala College for the constant encouragement and useful suggestions. We are grateful to Mr. Cris Thomas, Media Centre, Nirmala College.

We are thankful to Dr. K.R. Gupta, Honorary Advisor, Atlantic Publishers and Distributors (P) Ltd. and Mr. Harjeet Singh Publishing Co-ordinator, for publishing this book in a short time.

Thomas Varghese
K.M. Balakrishna

Contents

1

Introduction

The word nanotechnology is an umbrella term and it accommodates conventional physics, biology, chemistry, materials science and all engineering disciplines. This new concept in manufacturing makes most products with unique characteristics, such as cleaner, precise, lighter, stronger and less expensive. Nanotechnology will profoundly affect economy and society, as much as the industrial revolution has. Over the past few years, we have witnessed rapid advances in the field of nanotechnology on many fronts including materials, electronic devices, medicine and healthcare, biotechnology, biosensors, energy storage and information technology. These advances have led to the availability of an array of technologies for potential applications. Applications of nanotechnology during the next few decades can result a very large increase in computer speed, enormous storage capacity, efficient battery storage, reduction in the cost of pure water, stainless clothes and therapies for different types of ailments.

The twenty-first century demands the technology for miniaturization of devices into nanometer sizes while their performance is amazingly enhanced and is considered that nanotechnology will be the next revolution. The nanotechnology products, nanomaterials and their applications are still in developing stage and true revolution is years away, a few or many years. This technology aims tinier and faster instead of bigger and slower. Thus, nanotechnology provides access to the world of the smallest things. The benefits of nanotechnology are almost unlimited, however, they will be understood only if the evil effects of nanotechnology are studied and managed.

1.1 Definition

The word "nano" originates from the Greek word "nanos", which means dwarf. A nanometer (nm) equals 10^{-9} meter. Figure 1.1 shows that one nanometer is around the space occupied by 3-4 atoms placed end-to-end. Following examples help to create a sense of nano scaled objects: (i) diameter of one human hair is about 80,000 nm, (ii) size of the head of a pin is about 10^6 nm, (iii) size of an atom is about 1 nm, (iv) size of a DNA molecule is about 2.5 nm and (v) thickness of a red blood cell is about 5,000 nm. A nanoelement can be compared to a soccer ball, like a soccer ball to the earth. Figure 1.2 illustrates different objects from macroscale to nanoscale, and suitable models at different scales.[1] The figure 1.2 also shows appropriate tools and models for the study of objects at various sizes. In order to commemorate Feynman's great contributions to nanotechnology, it is suitable to name the nanometer scale as 'the Feynman (φ nman) scale'.

1 Feynman [φ] = 1 Nanometer [nm] = 10^{-9} meters = 10^{-3} microns [μ] = 10 Angstroms [Å]

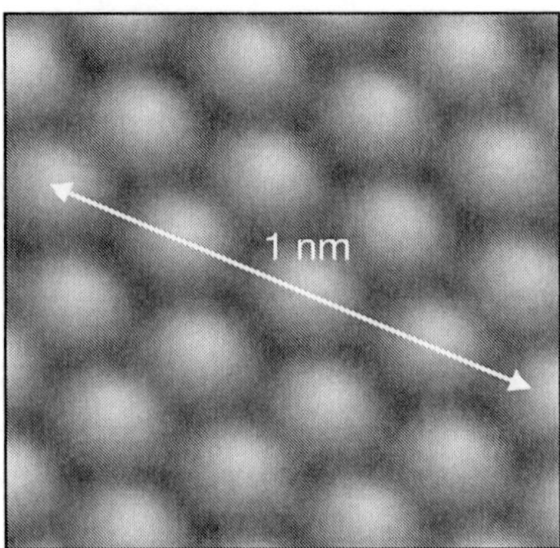

Fig. 1.1: Scanning tunnelling microscope image of carbon atoms on the surface of pyrolytic graphite.

Source: www.nanohub.org

Nanoscience involves researches to investigate new behaviours and properties of materials with nanoscale dimensions. Nanotechnology is the research and development of materials, devices, systems and products by manipulating shape and size at the nanometer scale with at least one novel property. The United States National Science Foundation [NSF] defines nanoscience or nanotechnology as studies that deal with materials and systems having the following key properties: (i) Dimension—at least one dimension from 1 to 100 nm, (ii) Process—designed with methodologies that shows fundamental control over the physical and chemical attributes of molecular-scale structures, and (iii) Building block property—they can be combined to form larger structures.[2] The national nanotechnology initiative of NSF defines nanotechnology as the understanding and control of matter at dimensions of 1 to 100 nm, where unique phenomena enable novel applications. In brief, nanotechnology is the ability to construct nanostructured materials and devices with atomic level precision.

1.2 Importance of the Nanoscale

The significance of the nanotechnology is mainly due to the fact that nanomaterials show properties, such as physical, chemical and electronic, which are rather different from their bulk counterparts. Some of the properties lie between that of the nano elements from which they can be composed of and those of the macroscopic materials. For similar applications, nanomaterials have superior performance properties than macroscopic materials. It is interesting that composites made from nanoparticles of ceramics or metals are much stronger than that predicted by theoretical models. For example, materials with a grain size of about 10 nm are found to exhibit enhanced mechanical properties as compared to their ordinary counterparts with large grain sizes. Thus, nanoscale is a magical point at which materials exhibit superior properties.

The invention of advanced microscopic instruments enhances the ability to construct nanostructures, which can help to explore new physical, chemical and biomedical properties of systems. The nanoscale is highly significant due to many reasons as follows:

(i) The quantum mechanical effects become significant in the nano regime. The fundamental properties of materials can be modified by designing materials at nanoscale dimensions.

(ii) Nanobiotechnology offers devices with bio-recognition properties, which help to study biochemical processes and to manipulate living cells at single molecule level. The nanotechnology has great promising applications in general medicine, tissue engineering, pharmacogenomics and surgery.

(iii) Nanoscale materials have large surface area to volume ratio, which make them suitable for use in composite materials, reacting systems, drug delivery systems and energy storage. Atom is very close to the surface or interface, and hence their behaviour at these higher-energy sites have a great influence on the properties of the material. For example, the reactivity of a metal catalyst particle generally increases appreciably as its size is reduced. Gold is chemically inert at macroscopic scale. However, gold becomes reactive and catalytic at nanoscale, and even melts at a lower temperature. The larger surface area permits simultaneous interaction of chemicals with the catalyst, which makes the catalyst more effective.

(iv) Large systems made up of nanostructures can have much higher density and electrical conductivity. This may result in new electronic device concepts, such as smaller and faster circuits, new sophisticated functions and low power consumption.

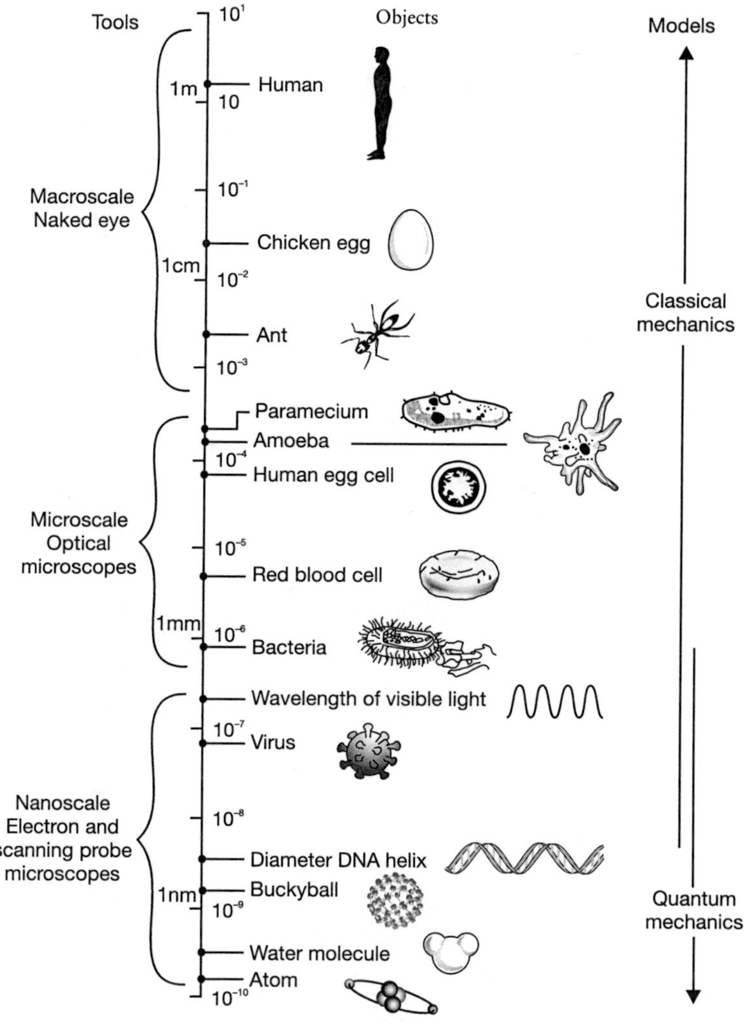

Fig. 1.2: Size comparisons from macroscale objects to nanoscale objects.[1]

Researchers hope to imitate nature's secrets of building from the nanoscale, to generate new processes and devices. For example, they have developed techniques to create water repellent surfaces by copying the nanostructure of lotus leaves. It is worth to note that many important functions of living organisms take

The new generation scientific tools that operate in nanoscale enable to collect data and to manipulate atoms and molecules on a very small scale. With these tools, it is found that many familiar materials act differently and have different characteristics and properties when they are in nanoscale quantities. Moreover, nanoscale materials can exhibit astounding characteristic properties that are not found at macroscopic scales. For instance:

- Graphite, a form of carbon, is soft and malleable. But carbon nanotubes exhibit tensile strength 100 times than that of steel.
- Samples of gold nanoparticles can appear in different colours, such as orange, purple, red or greenish, depending upon the size of the particles making up the sample.
- Nanoscale zinc oxide is transparent; however, at large scale it becomes white and opaque.
- Nanoscale aluminum can combust spontaneously and can be used in rocket fuel.
- At room temperature, nanoscale copper becomes a highly elastic metal and can be stretched up to 50 times its length.

place at the nanoscale. Natural nanoscale materials like proteins and other molecules control many functions and processes of the living systems. For example, haemoglobin is about 5 nm in diameter, which carries oxygen through the bloodstream.

1.3 History of Nanotechnology

Ancient people have fortuitously employed nanotechnology for thousands of years, but it is not clear, when they first began to use the advantage of nanophase materials. In the fourth century, Roman glass workers were able to fabricate glasses containing nano metals. A cup, called Lycurgus cup (depicts the death of King Lycurgus) made during this period is exhibited at the British Museum in London. It is made from soda lime glass containing nanoparticles of silver and gold. Its colour changes from green to red when a source of light is placed inside the cup. The glass windows of medieval churches also contain metal nanoparticles, which contribute to their beautiful colours. Ancient stained glass makers could produce different colours in stained glass windows by adding small amounts of gold and silver nanoparticles in the glass as shown in figure 1.3. The figure shows that they were able to produce various colours by varying size and shape of the nanoparticles. The shiny decorative surface, luster, found on some medieval pottery is due to the dispersion of spherical shaped metal nanoparticles. They used some simple techniques to produce these materials and are not fully understood even now.

In 1661, Irish chemist Robert Boyle questioned Aristotle's belief that matter is composed of earth, fire, water and air. But he proposed that tiny material particles combine in different ways to generate corpuscles. Peter Paufler and co-workers at the Technical University, Dresden, Germany studied samples from the blade of the Damascus sword made in the 17th century. Damascus sword was famous for their strength and sharpness. They investigated the presence of nanowires and nanotubes in the sample.

In 1857, Michael Faraday published an article in Philosophical Transactions of the Royal Society, which discussed about how the colour of glass windows of churches changes with metal particles. He identified the essential nature of nanoscale metal particles. The gold nanoparticles prepared by Faraday are preserved in the Royal Institution, London. In the German journal *Annalen der Physik* (1908), Gustav Mie discussed the variation of the colour of glasses with the size of the metal particles. James Clerk Maxwell (1867) mentioned some of the distinguishing concepts in nanotechnology and proposed a tiny entity called "Maxwell's Demon". He also produced the first colour photograph that depends on production of light sensitive silver nanoparticles.

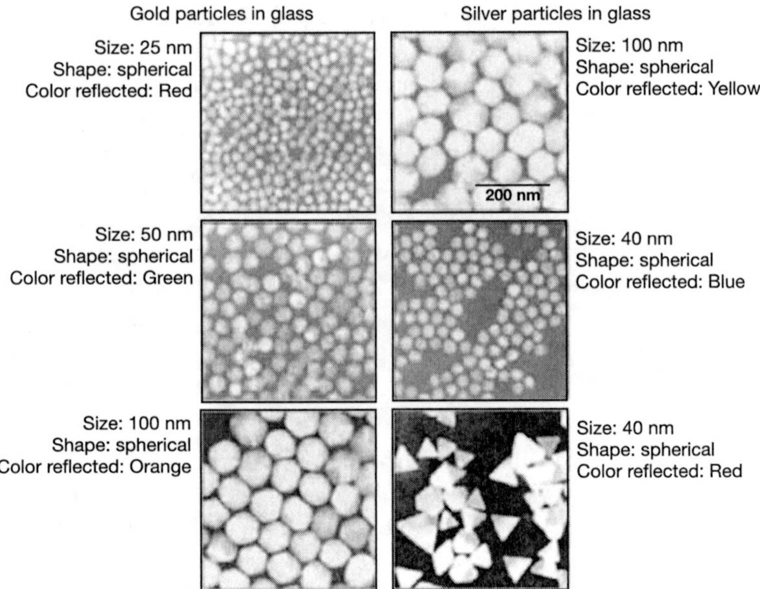

Fig. 1.3: Glass containing gold and silver nanoparticles of different sizes and shapes exhibiting different colours.

Source: Chad A. Mirkin, Institute of Nanotechnology, Northwestern University.

Chemical catalysis is an example of "old nanotechnology". Today, catalysts can speed up many chemical transformations, such as conversion of crude oil into gasoline, small organic materials into drugs and polymers and graphite into synthetic diamond. Most catalysts were investigated by trial and error method.

Peoples have been working with nanomaterials for centuries even without knowing the structure of nanoparticles. The discovery of high resolution microscopes helps them to see what they are working with. During the first decade of 20th century, Richard Adolf Zsigmondy made the first observations and size measurements of gold and other nanoparticles (~10 nm) by using ultramicroscope with dark field method. He was the first researcher who used nanometer as 1/1,000,000 of millimeter. He won the Nobel Prize in chemistry in 1925 for his work. There have been many important developments in characterizing nanomaterials during 20th century. Langmuir and Blodgett (1920s) introduced the idea of a monolayer material. Langmuir was awarded Nobel Prize in chemistry in 1932 for his contributions to monolayer materials.

The perception of manipulation of matter with atomic level precision was first introduced by Richard P. Feynman[3] in his speech entitled, "There's Plenty of Room at the Bottom—An Invitation to Enter a New Field of Physics". He delivered this amazing lecture on the occasion of the annual meeting of the American Physical Society on 29th December, 1959. It was published as an article in *Engineering and Science* magazine, California Institute of Technology in 1960. In his lecture, he envisioned the possibility and potential of nanotechnology.

> "A biological system can be exceedingly small. Many of the cells are very tiny, but they are very active; they manufacture various substances; they walk around; they wiggle; and they do all kinds of marvelous things—all on a very small scale. Also, they store information. Consider the possibility that we too can make a thing very small which does what we want—that we can manufacture an object that maneuvers at that level."
> —Richard P. Feynman, 1959

Feynman in his legendary talk described how to explore remarkable properties of materials by manipulating them at the molecular scale. Feynman presented fundamental concepts about miniaturization of printed matter, circuits and devices. "There's no question that there is enough room on the head of a pin to put all of the Encyclopedia Britanica", he quoted. He also predicted that a library with all the world's books would fit in a pamphlet in our hand. Almost all of the speculations made by Feynman in his lecture are under investigation by nanotechnologist's worldwide and his fictions have become reality today. But, his speculations were not understood by scientists at the time. Richard P. Feynman received the Nobel Prize in physics in 1965 for his work on quantum electrodynamics.

In 1974, Nario Taniguchi used the term "nanotechnology" to describe materials with scales less than a micrometer.[4] Eric Drexler has given widespread propaganda in popularizing nanotechnology.[5-7] In his book, *Engines of Creation* (1986), Drexler introduced a world of tiny machines or assemblers that can construct new structures with atomic level precision. If it is technically possible to achieve atom-by-atom construction of larger objects, it can be a new way of material synthesis and the result will be a second industrial revolution with more societal impacts.

Ralph Landauer (1957) introduced the concept of nanoscale electronics and the role of quantum mechanical effects on such devices. Alfred Cho and John Arthur (1968) discovered Molecular Beam Epitaxy (MBE), and thus improved the controlled deposition of single atomic layers. Gerd Binnig and Heinrich Rohrer (1981) investigated the Scanning Tunnelling Microscope (STM), and they were awarded Nobel Prize in 1986 for this work. STM images give the position of individual surface atoms. Richard E. Smalley and his co-workers discovered buckminsterfullerene (buckyballs) in 1985, which are soccer ball shaped molecules made up of carbon. In 1996, they were awarded Nobel Prize in Chemistry. Buckyball is the man-made allotropic form of pure carbon. Sumio Iijima working for NEC Corporation, Japan, discovered carbon nanotubes in 1991, while researching buckyballs using an electron microscope.[8]

The Dip Pen Nanolithography (DPN) using Atomic Force Microscope (AFM), the manipulation and positioning of single atoms with STM, and the trapping of single (3 nm diameter) colloidal particles from solution are some of the significant achievements of nanotechnology which Feynman forecasted in his talk. A few examples of nanomaterials are shown in figure 1.4.

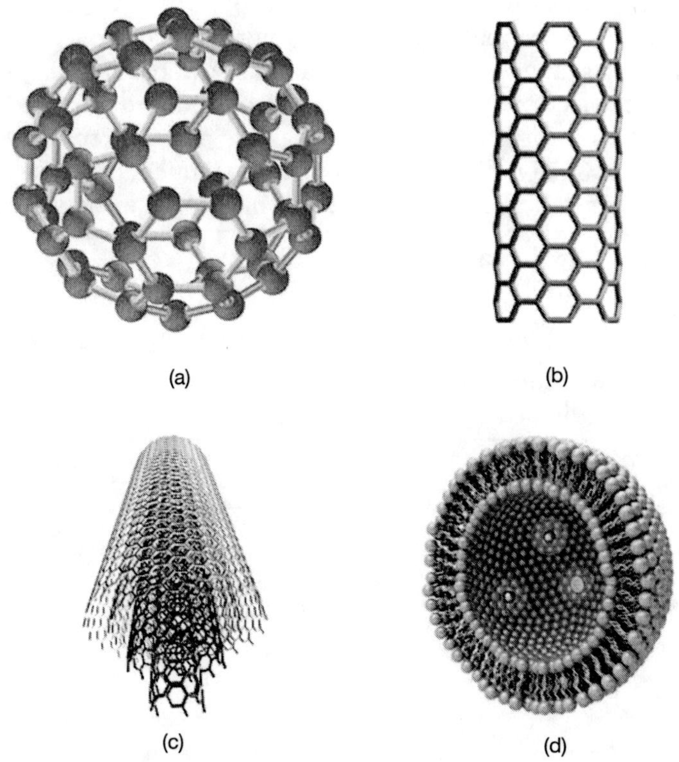

Fig. 1.4: Examples of nanomaterials: (a) Buckyball, (b) SWNT, (c) MWNT and (d) Nanoshell.

Sources: www.commons.wikimedia.org; www.robaid.com; www.techtransfer.universityofcalifornia.edu.

1.3.1 Moore's Law

The top-down approach to microelectronics synthesis appears to be governed by exponential time dependence. This exponential behaviour was first observed by Gordon E. Moore (1965) in his paper "Cramming more components onto integrated circuits". He also made an amazing forecast that the number of transistors made on a chip of given area would double in every two years.[9,10] His prediction is known as Moore's law. It has been observed that the transistor count in integrated circuits double in every two years as shown in Moore's law plot (figure 1.5). It is important to note that this trend has not only continued so far but has crossed the limit of the prediction. As a result, the size of individual electronic components will be reduced. In modern circuitry, size of electronic components reduces to tens of nanometers (~ 25 nm). Moore's law plot of number of transistors on an integrated circuit versus year is illustrated in figure 1.5.

Moore's Law Equation

Computer processing power in future years, $P_n = P_o \times 2^n$,

where P_o = computer processing power in the beginning year,

n = number of years to develop a new microprocessor divided by 2.

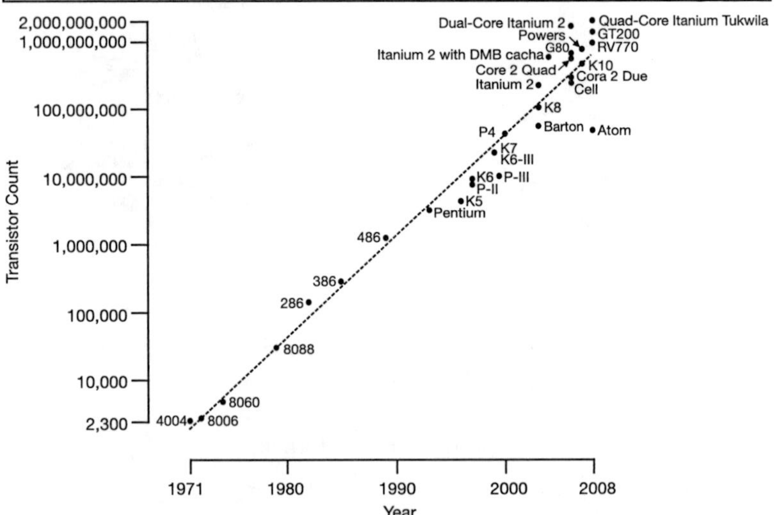

Fig. 1.5: Moore's law plot of number of transistor on an IC/CPU versus year.

Source: www.upload.wikimedia.org

In the last century, the transition from one technology to another has occurred several times in information industry. For example, the mechanical relay was replaced by the vacuum tube, which was then substituted by the transistor. Subsequently, the transistor gave way to the current integrated circuit.

1.4 Applications

The invention of precise microscopic instruments with improved capabilities to understand and manipulate materials at the nanoscale promises a number of elegant new technologies in future. Dynamic improvements will take place in varied fields such as medicine, drug delivery, communications, quantum computing, data storage, energy storage and robotics.[11-13] All these changes may cause tremendous changes in the existing consumer products and industries. An overview of some of the promising applications of nanoscience and nanotechnology is given in this section. A detailed discussion of the various foreseeable applications of nanotechnology is given in chapter 7.

1.4.1 Medicine

Nanomaterials exhibit unique properties like size and shape dependent optical properties, large surface area-to-volume ratio, high surface energy and tunable surface properties. These novel properties make them suitable for bio-sensing applications.[1] A wide variety of nanomaterials have found very useful applications in many kinds of biosensors for the diagnosis of diseases, drug delivery, cellular imaging and so on.[14] Usually physical disorder of molecules and cells cause most of the diseases, and hence medicine at molecular level can cure ailments.

Nanotechnology offers wide scope in medicine. Assembler-based manufacturing will enable nano tools for molecular-scale surgery to repair and reorganize cells. It is worth mention that medicinal fluids containing nanorobots can be programmed to perform delicate surgeries, repair cancer cells and mutations in DNA, destroy toxic chemicals and attack viruses to make them harmless. Also, nanotechnology tools will help to make biocompatible joint replacements, which will last for the entire life of the patient. Moreover, nanotechnology will help to improve health and physical capabilities of human beings.

With faster and cheaper diagnostic equipments, better diagnostic tests will be conducted. For example, DNA mapping of the newborn children may help to point out future potential problems and thereby prevent disease before it takes hold. The dangerous side effects of chemotherapy and other treatments are mainly due to inefficient drug delivery systems, which cannot locate their target cells accurately. Researchers at Harvard University have reported that it is possible to attach special RNA strands to nanoparticles filled with chemotherapy drug. Since the RNA strands are attracted to cancer cells, nanoparticles adhere to cancer cells and release the drug into them. This type of drug delivery will produce only fewer side effects than those produced by usual chemotherapy.

1.4.2 Nanoelectronics

Nanotechnology has great impact on electronics engineering and industry. It has wide applications in every sector of electronics field, which includes field emitters and flat panel displays, molecular switches, nanotube transistors and field effect transistors, nanotube ICs, biomedical electronic devices and nano-biosensors.[2]

1.4.3 Batteries

Lightweight and high energy density batteries are of great demand because of its widespread use in portable electronic instruments such as mobile phones, navigation devices and laptop computers. Sol-gel based nanosynthesis provides materials with aerogel structure and are best suited for separator plates in batteries because they can hold considerably more energy than conventional batteries. For example, nanoparticle based Nickel-metal hydride batteries require less frequent recharge.

1.4.4 Environmental Protection

Nanotechnology has tremendous impact on environmental protection, such as pollution treatment and remediation in which waste atoms can be recycled or they can be kept under proper control.[1] The potential benefits of nanotechnology in environmental protection include detection, sensing and eradication of contaminates from air, water and soil, and creation of ecofriendly industrial process, which reduces waste products. For example, iron nanoparticles can be used to remove chemicals in groundwater because they react to chemicals more efficiently as compared to larger iron particles. Programmable airborne nanorobots can help to rebuild the thinning ozone layer in the atmosphere. They can also be used to remove excess CO_2 in the atmosphere.

1.4.5 Food and Agriculture

Nanotechnology will provide new techniques for water filtration as well as desalination, which will be more economical.[1] It can also help to improve agricultural yields, and will be much greater than before. This new technology has to do much in food industry also. For example, nanotechnology will help to develop new functional materials and design new instruments for food preservation and bio-security. Bayer Company has marketed an airtight plastic packaging using nanotechnology to preserve food materials fresher. Nanotechnology can also help to modify the genetic constitution of the crop plants, thereby helping improvement of crop plants. Nanotechnology based plant disease diagnostics help to detect exact strain of virus and stage of application of some therapeutics to stop the disease.

1.4.6 Energy

Energy storage, energy conversion, energy saving and effective utilization of renewable energy sources are some of the important energy applications of nanotechnology. Conventional solar cells have low efficiency of light conversion. But, nanotechnology can help to improve their efficiency using nanostructures with a continuum of band gaps. Power consumption for lighting can be reduced considerably by using nanomaterials based light emitting diodes (LEDs) and quantum caged atoms (QCAs). Hydrogen based fuel cells are eco-friendly form of energy because their byproduct is water. Nanomaterials such as nanotubes, zeolites or alanates have a large number of nanosized pores and hence, such materials are suitable for hydrogen storage. Usually, nanomaterials used in fuel cells are catalysts containing carbon based metal particles with diameters in the range 1-5 nm.

1.4.7 Nanomaterial-based Products

The invention of new microscopic techniques improved the ability to see and manipulate nanosized materials. This will result in an array of possibilities in scientific and industrial applications. Moreover, nanotechnology can revolutionize products everywhere, producing a world of new products and devices. Nanotechnology can modify the nature and properties of manufactured products and hence, its impact is much more than the semiconductor revolution. Amazingly, a very large variety of nanomaterial-based commercial products are available in the market.

It is important to note that the performance of the conventional materials can be improved by adding suitable nanoparticles. Manufacturers use this advantage to change the properties of materials. For example, some cloth manufacturers use nanosized whiskers in the fabric to make water and stain resistant fabric. Companies are now manufacturing different nanoparticles for use in commercial products, such as stain free cloths, crack free paints and self-cleaning windows.

Some examples of nano products/devices:

- Bandages coated with silver nano-crystals are designed by Smith & Nephew to prevent infection. Exploiting the anti-bacterial properties of nano silver, a large number of products are marketed. For example, sheets, towels, appliances, socks, toothbrushes, toothpastes and children's toys.
- Nanocosmetics: Nanosomes are nanoparticles, which can penetrate the outer layer of the skin and deliver nutrient materials encapsulated in them to cell below the outer layers. They can move to deeper layers and the cosmetic formulations are effectively absorbed and new cells are formed. These new cells help the skin to remain soft and wrinkle free regardless of age.
- Titanium dioxide (TiO_2) nanoparticles are transparent and block ultraviolet (UV) light. This property of TiO_2 nanoparticles can be used in sunscreens and in plastic food wraps for UV protection.
- Hydroxyapatite nanoparticles possess similar chemical structure as tooth enamel. To provide anti-bacterial

coating on teeth, researchers are hoping to incorporate the nanoparticles in toothpaste. Sangi Co. Ltd. (Japan) has been marketing a nano-hydroxyapatite based toothpaste since 1980.

- Stain Defender fabrics with a molecular coating is another significant product marketed by Nano-Tex, that adheres to cotton fibre, forming an impenetrable barrier that causes liquids to bead and roll off.
- Self-cleaning window glass covered with a layer of TiO_2 nanoparticles is marketed by Pilkington. The dirt on the surface of the glass is loosened as the nanoparticles interact with UV rays from sunlight, and the dirt is washed off when it rains.
- Nanoscale synthetic carotenoids are marketed by BASF, as a food additive in lemonade, fruit juices and margarine. The carotenoids are antioxidants that can be converted to Vitamin A in the body and are easily absorbed.
- Syngenta Agrochemical Corporation sells nanoparticles based pesticide products. The company claims that the chemical is directly absorbed into the plant's systems and cannot be washed off by rain or irrigation.
- Altair Nanotechnologies proposed a water-cleaning product for swimming pools and fishponds, which incorporates nanoparticles of a lanthanum-based compound that absorbs phosphates from the water and prevents algae growth.
- A variety of nanobiosensors are manufactured by various companies, such as ACLARA, Agilent Technologies, Calipertech and I-STAT.

1.5 Risks of Nanomaterials

Nanotechnology has a significant impact on science and society, and all sectors of economy. However, they may cause new risks to human health. In health sector, the most significant concerns are likely to be the safe and ethical use of nanomaterials. It is very difficult to detect and control nanoparticles due to their extremely small size. Researchers, workers, consumers or patients may accidentally inhale nanoparticles. The models and predictability of their molecular interactions are not yet known. Thus, precautions to avoid inhalation are needed. Usual safety precautions may not provide adequate protection, creating a need for new and comprehensive evaluation research studies. Only a few research studies are reported about the safety of nanomaterials.

Some studies on animals reported that inhaled nanomaterials enter into the brain via olfactory neurons or from lungs to the blood stream. Once they enter the body, they become more chemically reactive due to their larger surface area to volume ratio and hence, interact with biological molecules.[15,16] The interaction with cell DNA causes inflammation and oxidative damage. Engineered modifications to nanomaterials, (e.g., surface coatings), can modify a material's solubility, reactivity, toxicity and other properties, and thereby reduce the risks associated with a material early in its design. The nanosize air pollutants would result in lung cancer and cardiopulmonary diseases. The risk is high among individuals with pre-existing heart and lung ailments. Besides these, nanoparticles may cause oxidative stress in brain, skin cells, liver and lung tissues. Obviously, guidelines along with the risk possible with nanomaterials should be generated to preserve human dignity and integrity.

Nanoresearchers and technologists have published a few challenges to ensure the safe handling of nanoparticles. They are: (i) develop suitable instruments to evaluate exposure to engineered nanomaterials in air and water, (ii) develop appropriate methods to evaluate the toxicity of nanomaterials, (iii) develop proper models for predicting the potential impact of nanomaterials on human health and (iv) develop strategic programmes that facilitate relevant risk focused research.

REFERENCES

1. John Mongillo, *Nanotechnology*, 101 (Pentagon Press, 2009).
2. M.C. Roco, S. Williams and P. Alivisatos (Eds.), "Nanotechnology Research Directions: IWGN Workshop Report—Vision for Nanotechnology R&D in the Next Decade", WTEC, Loyola College in Maryland, September (1999).

3. R.P. Feynman, *Engineering and Science* magazine of Cal. Inst. of Tech., 23, 22 (1960).
4. N. Taniguchi, *Proc. Intl Conf. Prod. Engg.*, Tokyo, Part II (Jap. Soc. Precision Engineering) (1974).
5. Drexler, K.E., *Nanosystems: Molecular Machinery, Manufacturing, and Computation* (John Wiley & Sons, Inc.: NY, 1992).
6. Drexler, K.E., *Proc. Natl Acad. Sci.*, USA 78 5275-78 (1981).
7. Drexler, K.E., *Engines of Creation* (Anchor Books: New York, 1986).
8. Iijima, S., *Nature*, 354, 56 (1991).
9. Moore, G., *Electronics*, 38, No. 8 (1965).
10. Moore, G., *IEDM Technical Digest*, 11 (1975).
11. Richard Booker and Earl Boysen, *Nanotechnology* (Wiley, 2005).
12. Poole, C.P. and Owens, F.J., *Introduction to Nanotechnology* (John Wiley & Sons, 2006).
13. Bhusion, B., *Handbook of Nanotechnology* (Springer—Heidelberg: NY, 2004).
14. Chiu, T. and Huang, C. (2009) Aptamer functionalized nanobiosensors. *Sensors* 9: 10356-88.
15. Peters, A., Dockery, D.W., Heinrich, J. and Wichmann, H.E., *European Respiratory Journal*, Vol. 10, No. 4, 872-79 (1997).
16. Borm, P.J.A., *Particle and Fibre Toxicology* 3, 11 (2006).

2

Nanomaterials

2.1 Introduction

Nanomaterials are materials with structural units on a nanometer scale in at least one dimension. They are metals, ceramics, polymeric materials, or composite materials with very small dimensions in the range 1-100 nanometers (nm). According to Publicly Available Specification (PAS), published by the British Standards Institute, a nanomaterial is defined as a material having one or more external dimensions in the nanoscale or which is nanostructured. Nanomaterials exhibit unique physical and chemical properties such as electrical, magnetic, mechanical, catalytic, electronic and thermal properies. These novel properties make them suitable for applications in medical, military, scientific, electronic, environmental and commercial sectors. The development of nanotechnology has been stimulated by enhancement of tools like electron microscopy and scanning tunnelling microscopy, to see the nanoworld. This chapter gives an overview of various types of nanoscale materials, their properties and significant foreseeable applications.

Nanomaterials are broadly classified into two: nanostructured materials and nanophase materials (or nanoparticles).[1-5] Nanostructured materials are condensed bulk materials, which are made of nanoscale grains. However, nanophase materials are usually nanoparticles distributed uniformly in a medium. It is interesting that there are other terms commonly used in describing different nanoparticles. Nanocrystals are specifically denoted to single crystal nanoparticles. Very small nanoparticles are often called quantum dots. They have nano-dimensions in all the three directions. Quantum dots are used to describe small particles that exhibit quantum size effects. Similarly, quantum wires are referred to as nanowires when exhibiting quantum effects.

To distinguish nanomaterials from bulk, it is important to demonstrate the unique properties of nanomaterials and their prospective impacts in science and technology. Nanomaterials exhibit novel or superior size dependent properties as compared to their bulk counterpart. For instance, nanoscale TiO_2 and ZnO are transparent and they can absorb and reflect ultraviolet light. There are various types of man-made nanomaterials, such as metals, metal oxides, semiconducting materials, nanofilms, nanosheets, nanowalls, carbon based nanomaterials, inorganic nanotubes, nanowires, and quantum dots, and many others are expected to investigate in future.[6-10] Moreover, a large number of nanomaterials with an array of useful properties make them ideal for variety promising applications.

2.2 Carbon-based Materials

Carbon based nanomaterials mainly consist of carbon, and are generally in the form of hollow spheres, ellipsoids or tubes. Carbon nanomaterials with spherical or ellipsoidal structures are called fullerenes, and cylindrical structures are referred to as nanotubes.[11-13] These materials have a range of properties and a large number of possible applications.

2.2.1 Fullerenes

Fullerenes consist of hexagonal and pentagonal rings of carbon atoms with spherical or ellipsoidal and tube shapes. In the mid-1980s a new allotropic form of carbon was invented by Kroto and Smalley, carbon 60 (C_{60}). It is called "buckminsterfullerene", in respect of the renowned architect Buckminster Fuller who made geodesic domes. Any closed carbon cage is then named as fullerene.

The fullerene C_{60} has spherical molecular configuration with a diameter of about 1 nm. It consists of 60 carbon atoms, which are arranged in 20 hexagons and 12 pentagons, similar to the soccer ball configuration. Spherical fullerenes are called buckyballs. The smallest buckyball is C_{20} and it consists of 12 pentagons only. Bigger buckyballs are denoted as C_{2n}, where n = 12; 13; 14..., and have non-isomorphic shapes. Models of C_{60} and C_{540} are shown in figure 2.1.

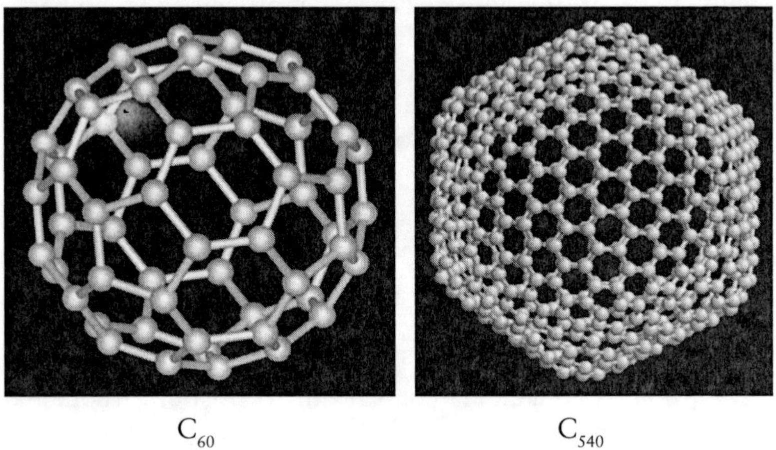

Fig. 2.1: Fullerenes C_{60} and C_{540}.
Sources: www.shungite.com; www.static.newworldencyclopedia.org

Fullerenes generally exhibit properties such as superconductivity, heat resistance, chemically unreactive, soluble in many solvents and so on. Fullerenes have many potential applications which include, drug delivery vehicle and miniature ball bearings to lubricate surfaces. Structure, properties and applications of fullerenes are presented in chapter 6.

2.2.2 Carbon Nanotubes (CNTs)

CNTs were discovered by Iijima (1991). They are cylindrical fullerenes. Single walled nanotube (SWNT) and multi walled nanotube (MWNT) are the two types of CNTs. SWNT consists of one cylindrical tube with a diameter of about 1-2 nm, while MWNT consists of several concentric tubes having a diameter of about 2-25 nm. The interlayer distance of MWNT is around 3.3 Å. Both the CNTs have a length range from several micrometres to centimeters and their length-to-diameter ratio can go beyond 10,000. Models of SWNT and MWNT carbon nanotubes are shown in figure 2.2.

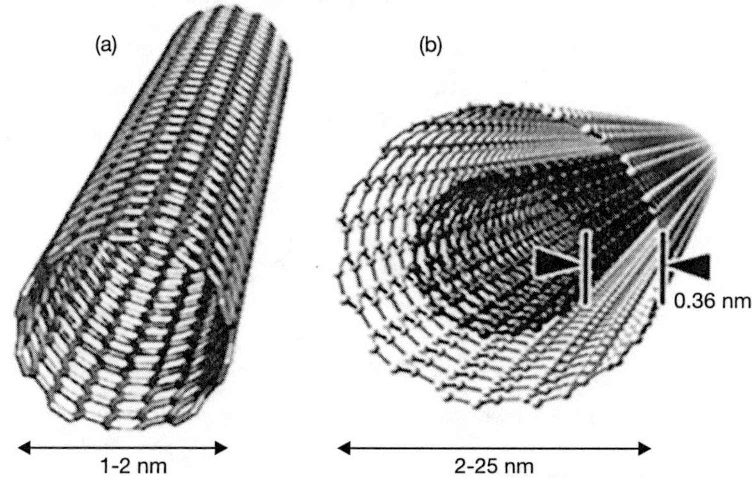

Fig. 2.2: Models of carbon nanotube: (a) SWNT and (b) MWNT.

Source: www.files.danmengshuai.webnode.com

CNTs have a significant role in nanotechnology, because they exhibit unique chemical and physical properties. They show very high mechanical properties. For example, Young's modulus is above 1 TPa and tensile strength is over 126 GPa. Besides these, CNTs are flexible along their axis and are very good electrical conductors. Furthermore, all these properties make them ideal for a variety of potential applications, such as in microelectromechanical devices, AFM tips and display devices. A detailed discussion of carbon based nanomaterials and carbon nanotubes is given in chapter 6.

2.2.3 Nanobuds

Carbon nanobuds are fabricated by combining carbon nanotubes and fullerenes. In nanobuds fabrication, fullerene buds are covalently bonded to the outer sidewalls of the CNT, as shown schematically in figure 2.3. They possess amazing properties of both fullerenes and CNTs. They have excellent properties like huge capacity for hydrogen or lithium storage, catalytic reactivity and good field emitters. In composite nanobud materials, the bonded fullerene molecules may act as molecular anchors and thereby prevent slipping of the nanotubes, which may improve the composite's mechanical properties.

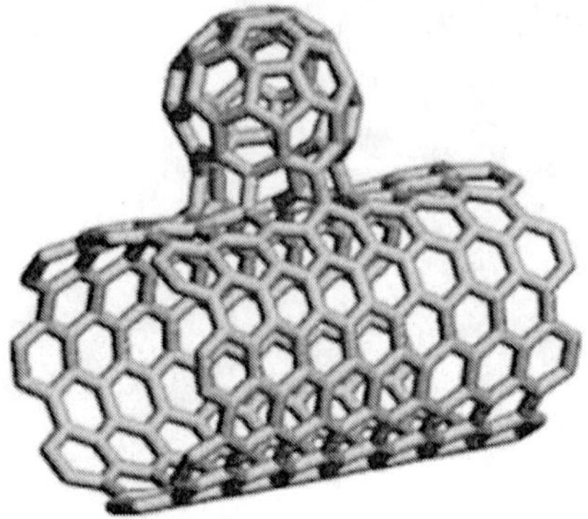

Fig. 2.3: Model of a nanobud.

Source: www.upload.wikimedia.org

2.3 Inorganic Nanotubes

Inorganic nanotubes and fullerene materials were discovered just after the discovery of carbon nanotubes. These materials have outstanding properties like effective resistance against shockwave impact, better catalytic activity and capacity for hydrogen and lithium storage. Inorganic nanotubes have number of potential applications due to their novel properties. For example, oxide based nanotubes (TiO_2) can be used in catalysis and energy storage.

2.4 Nanoshells

Nanoshells are a new type of nanoparticles and are ball shaped. They are made up of a core of nonconducting substance such as silica (glass), which is coated with an extremely thin metallic layer (shell) such as gold or silver.[14] Nanoshells are about the size of a virus or about 100 nm wide. The physical properties of gold nanoshells are similar to gold colloid. The absorption of light by gold nanoshells creates a red colour which can be used in medical products. It is worth mentioning that the optical properties of nanoshells change as the relative size of the core and the shell changes. Hence, it is possible to change the colour of nanoshells by varying the dimensions of core and shell, and can tune across a wide range of spectrum. Metal nanoshells were invented by the researchers at Rice Quantum Institute. The synthesis of nanoshells generally consists of following simple steps:

(i) Grow dielectric nanoparticles (e.g., silica) dispersed in solutions.

(ii) Bond small metal seed (~1-2 nm) colloid into the surface of the dielectric nanoparticles by molecular linkages. A discontinuous metal colloid layer is formed over the surface of the dielectric core.

(iii) Breed additional metal on the seed metal colloid adsorbates through chemical reduction in solution.

This approach is successful for the growth of both gold and silver nanoshells on silica nanoparticles. Figure 2.4 shows growth stages of gold shells over silica nanoparticles.

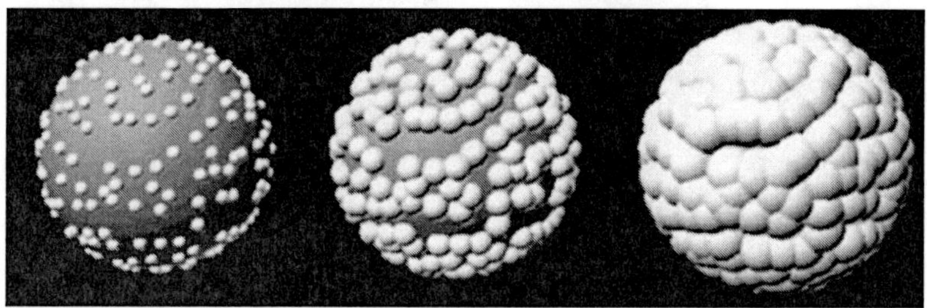

Fig. 2.4: Growth stages of gold shells around silica nanoparticles.

Source: www.rqi.rice.edu

Nanoshells have found potential application in cancer treatment without any toxic side effects as in chemotherapy. These nanoshells can be injected safely into the body. When the nanoshells in the body are illuminated with laser beam, intense heat will be produced and thus destroys the tumour cells. Nanoshells combined with lasers can also be used to eradicate cancer cells without damaging untargeted cells. Nanoshells are already being developed for other medical applications such as drug delivery and test for proteins associated with Alzheimer's disease.

2.5 Quantum Well

A quantum well is a thin layer in which particles (electrons or holes) are confined in the dimension perpendicular to the layer surface, while movement is possible in other dimensions. The confinement of particles is a quantum effect and it will occur as thickness of the quantum well approaches the deBroglie wavelength of charge carriers, resulting discrete energy levels to carriers. Quantum well structures in semiconductors are formed by sandwitching a material between other layers of another material having a wider band gap. For example, GaAs quantum well can be sandwitched between the layers of aluminium arsenide having a large bandwidth. The thickness of the quantum well is typically about 5-20 nm. These quantum well structures with monolayer thickness can be synthesised by MBE or metal organic CVD process. In laser diodes, semiconductor quantum wells are used in the active region. These quantum wells are embedded between two wider layers with a large band gap. The cladding

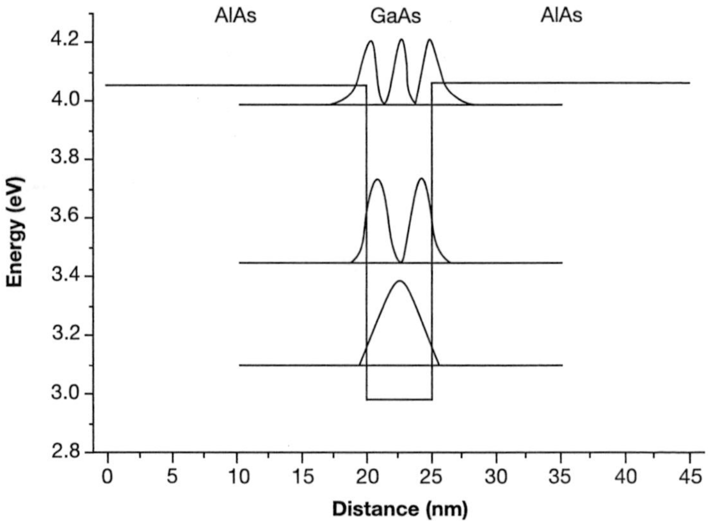

Fig. 2.5: A 5 nm GaAs quantum well.

Source: www.nextnano.de

layers act as a waveguide. But particles are captured by the quantum well as the difference in band gap energy is sufficiently large. A 5 nm GaAs quantum well embedded between AlAs layers is shown in figure 2.5.

The density states of electrons in quantum wells as a function of energy, and possess distinct steps due to quasi-two dimensional nature. Additionally, the reduced amount of active materials in quantum well leads to better performance in optical devices, such as laser diodes for DVDs and laser pointers, infra-red lasers in fibre optic transmitters and IR photodetectors for IR imaging. Quantum wells can also be used to develop High Electron Mobility Transistors (HEMTs), which are found applications in low-noise electronic devices.

2.6 Quantum Wires

Quantum wires are ultrathin wires or linear arrays of dots with a diameter of about 1-10 nm. They are defined as nanostructures with a diameter or thickness constrained to a few nanometers, but no restriction to length. Quantum mechanical effects are significant at these scales and hence, they are named quantum wires. They are formed by self-assembly of atoms on a solid surface. Quantum wires are made from metals (e.g., Ni, Pt, Au), semiconductors (e.g., Si, InP, GaN) and insulators (e.g., SiO_2, TiO_2). Molecular quantum wires are formed by repeated (organic or inorganic) molecular units (e.g., DNA). Quantum wires are also known as nanorod or nanowires. Quantum wires are electrically conducting wires, their transport properties are governed by quantum size effects. Because of the quantum confinement of free electrons across the wire, transverse energy is quantized into discrete energy values E_0, E_1, etc. It is important to note that the classical equation for electrical resistivity of a wire, $r = RA/l$, is not valid for quantum wires, where l is the length of the wire and A its cross-sectional area.

The preparation of nanowires is based on growth techniques, such as self-assembly processes, CVD and MBE. These fabrication techniques are discussed in chapter 4. The materials used for the growth of nanowires should be soluble in the catalyst nanoparticles. For example, gold catalyst nanoparticles are used to create silicon nanowires because silicon is soluble in gold. The nanowires, which are made from CNTs exhibit unique properties, such as high electrical conductivity and large tensile strength. Semiconductor nanowires have shown important optical, electronic and magnetic properties.

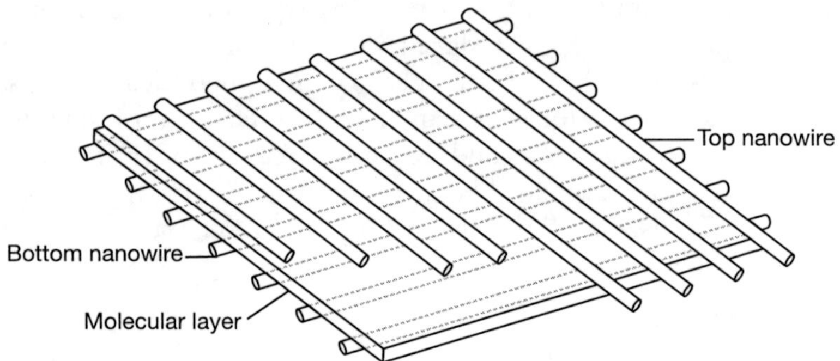

Fig. 2.6: Memory device using a crossbar array of nanowires.

Nanowires have an array of potential applications, which include high density data storage, various nanodevices and metallic interconnect. Researchers have reported that memory devices based on indium oxide nanowires can store 40 gigabits per square centimeter. For data stotage, each memory unit is formed at the intersection of two nanowires. Storage device using crossbar array of nanowires is shown in figure 2.6. Nanowire biosensors can be used to detect diseases in blood samples. In such biosensors, nanowires are the working parts and they are attached to some nucleic acid molecules. These molecules are attached to a cystic fibrosis gene in the blood sample, if it is present. As a consequence, there is a change in the conductance of the quantum wire and it may result a current flow. These types of sensors give immediate analysis of blood samples. Nanowires also have tremendous applications in optoelectronics, microelectronics, photonics and field emission devices.

2.7 Quantum Dots

Semiconductor nanoparticles and quantum dots were predicted in the 1970s, however, their experimental investigations were reported in the early 1980s. Quantum dots are semiconductors whose excitons are confined in three dimensions.[15] Hence, their properties lie between those of bulk semiconductors and those of discrete molecules. Quantum dots were invented by Louis E. Brus, and the term "Quantum Dot" was called by Mark Reed. The conductivity of quantum dots changes with the size and shape of individual crystals. The band gap is large for small crystals and hence, more energy is liberated when the crystal is deexcited. Since wavelength or colour of light depends on the energy liberated, the optical properties of the particles can be tuned by controlling their size. Moreover, quantum dots are made to absorb or emit specific colours by manipulating their size. It is important that high level control over the conductivity of the quantum dots is possible because the size of the crystals produced can be controlled. Monodispersed crystalline quantum dots with a dimension of about 2 nm were reported in literature. Atomic force microscopic image of InAs quantum dots is shown in figure 2.7.

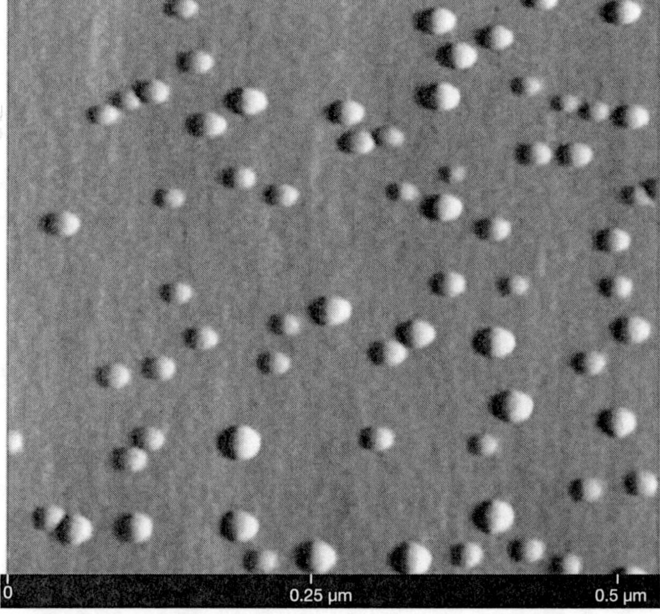

Fig. 2.7: A typical AFM image of InAs quantum dots.

Source: www.essential-research.com

The flow of electrons through quantum dots can be controlled by applying suitable small voltages to the leads, and thereby spin and other properties therein can be measured. Quantum dots have an array of promising applications in solar cells, lasers, medical imaging, biological labels, composites and quantum computation. Figure 2.8 illustrates density of states in different semiconductor structures, such as bulk, quantum well, quantum wire and quantum dot.

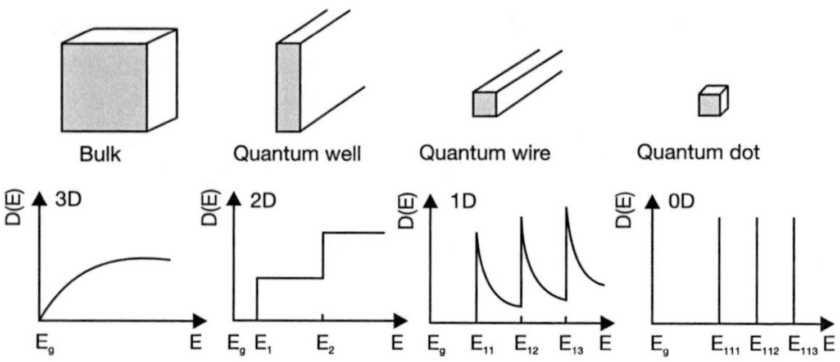

Fig. 2.8: Density of states in different semiconductor structures.
Source: www-opto.e-technik.uni-ulm.de

2.8 Dendrimers

A dendrimer is a macromolecule having 3D structure with superior surface functionality and versatility. The term "dendrimer" originated from two Greek words: *dendron,* means tree and *meros,* means part. There are different types of dendrimers, and their multivalent and monodisperse properties make them suitable for various significant applications, such as drug delivery, gene therapy, chemotherapy and environmental clean-up.

Dendrimers are created from an initial atom and other elements are bonded to it by a series of chemical reactions, generating a spherical branching structure. By repeating the process, successive layers will be added and the size of the dendrimer can be expanded to the desired level. An initiator core, interior layers (generations) composed of repeating units, attached to the interior core and exterior attached to the outermost interior generations are the main components of dendrimers. The dendrimer structure is shown in figure 2.9.

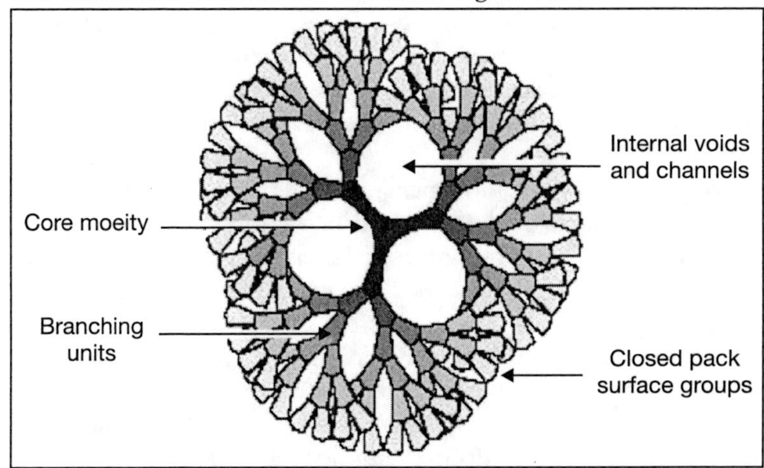

Fig. 2.9: The dendritic structure.

Source: www.inventi.in

Dendrimers have many potential applications, which are based on their properties like unparalled molecular uniformity, multifunctional surface and internal cavities. They can be used as drug carriers by holding drugs within their structure, or by interacting with medicines via electrostatic or covalent bonds. The modification of poly(lysine) dendrimers with sulfonated naphthyl groups help to use them as antiviral drugs against the herpes simplex virus. The excellent properties of dendrimers like branching, multi-valency and globular structure make them ideally fit for various pharmaceutical and diagnostic applications. Large numbers of drugs, which are being developed today, are facing problems like poor solubility, bioavailability and permeability. In such problematic drugs, dendrimers can be used as a tool for optimizing drug delivery.

2.9 Biological Nanomaterials

The study of biological nanomaterials can help to improve the production of synthetic nanomaterials with similar characteristics.[16] Biological systems contain many nanophase materials. For example, most of the living systems produce mineral materials (e.g., bone) with nanoscale particle structures. The biomineralization process consists of delicate biological control mechanisms that produce well defined nanomaterials. The study of these processes helps to construct novel materials. These materials can be modified by changing the biological situation in which they are produced. The biological materials can also be subjected to *in vitro* manipulation following extraction. Moreover, biological materials are used as the starting material for many of the standard procedures for the synthesis and processing of nanomaterials, such as vapour techniques, mechanical attrition, etc. For example, ferritins (a class of proteins) provide a biological system in which living organisms can synthesize and interact with nanosized particles of iron in the organism, such as oxyhydroxides and oxyphosphates. These proteins have been found in many types of living organism, from bacteria to man. The important biological functions of ferritins include the storage, transport and detoxification of nanosize iron in the organism.

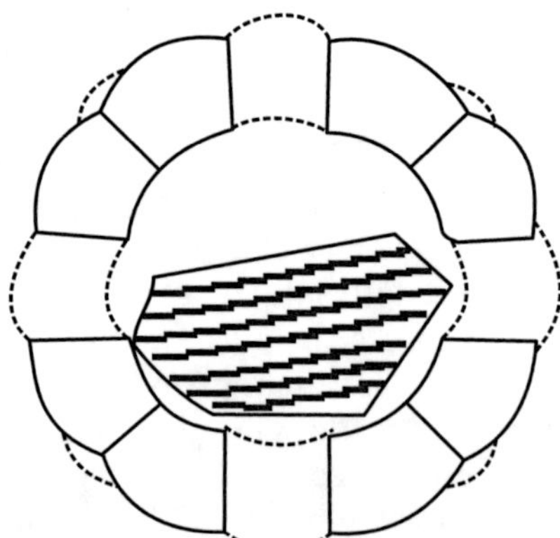

Fig. 2.10: Schematic diagram of a ferritin molecule.

The ferritin molecule which consists of a spherical protein shell, whose inner diameter is ~8 nm and outer diameter ~12 nm, is shown in figure 2.10. The spherical protein shell consists of 24 protein subunits and these subunits self-assemble to form a structure with channels, which permit iron ions to pass in and out. Ferritins provide a range of possibilities in the context of the synthesis and study of nanomaterials. Native ferritin consists of well separated nanometer sized magnetic particles, produced by biologically controlled mineralization within the cavity of the protein shell. The native material can provide a range of magnetic small particle model systems for the investigation of nanomaterial behaviour by a variety of techniques. The interpretation of this behaviour can of course assist in our understanding of the extracted material and, in turn, the system from which it has come.

2.10 Diamondoids

> "The constraints of predictability, stability and positional control in the face of thermal vibration all favour the use of diamondoid covalent structures in future nanomachines. Present-day experience with design and computational modeling of such devices has generated useful design heuristics, including reasons for favoring diamondoid materials with many noncarbon atoms."
>
> —K. Eric Drexler

Diamondoids are organic compounds with unique structures and unusual properties. Diamondoids are usually known as cage hydrocarbons. In 1933, diamondoids were discovered and liberated from Czechoslovakian petroleum, and they were found in many different crude oils, with varying concentrations and compositions. They possess diamond like fused ring structures, and are highly symmetrical and strain free. They are one of the best candidates for molecular building blocks (MBBs) to construct nanostructures because of their properties like rigidity, strength and assortment of their 3D shapes. Members of this polycyclic diamondoid family are adamantane, homologous adamantane, tria-, tetra-, penta- and hexamantane.[17,18]

Adamantanes have unusual physical and chemical properties due to their unique structures. The cage-like structure of adamantane can be used for the encapsulation of drugs or other compounds. Usually diamondoids are considered as saturated, polycyclic and cage-like hydrocarbons. The general molecular formula of homologous polymantane series is $C_{4n+6}H_{4n+12}$, where n = 1, 2, 3, ... (n = 1 for adamantane). Each higher diamondoids shows more structural complexity and different molecular geometries.

Figure 2.11 shows the chemical structures of adamantane, diamantane, triamantane and tetramantanes. The lower adamantologous, such as adamantane ($C_{10}H_{16}$), diamantane ($C_{14}H_{20}$) and triamantane ($C_{18}H_{24}$), have only one isomer. Higher polymantanes contain number of isomers and non-isomeric equivalents, depending upon the configuration of the adamantane units. The possible tetramantanes are iso-, anti- and skew-tetramantane. All these tetramantanes are isomeric. The iso-tetramantane contains three carbon atoms, but two quaternary carbon atoms in anti- and skew-tetramantanes. Six out of seven possible pentamantane structures are isomeric ($C_{26}H_{32}$), following the molecular formula of the homologous series. But the last pentamantane structure is non-isomeric ($C_{25}H_{30}$). There are twenty-four possible hexamantane structures. Seventeen of these structures are regular cata-condensed isomers ($C_{30}H_{36}$). But six hexamantane structures are irregular cata-condensed isomers ($C_{29}H_{34}$), and the remaining one is peri-condensed ($C_{26}H_{30}$).

It is important that adamantane crystallizes in f.c.c. lattice and thus the molecule is free from both angular and torsional strain. Adamantane shows only cubic and octahedral structures at the beginning of crystal growth and this unusual structure makes remarkable changes in physical properties. Melting point of adamantane is 269°C and it is the highest melting hydrocarbon known. Adamantane sublimes even at normal temperature and pressure.

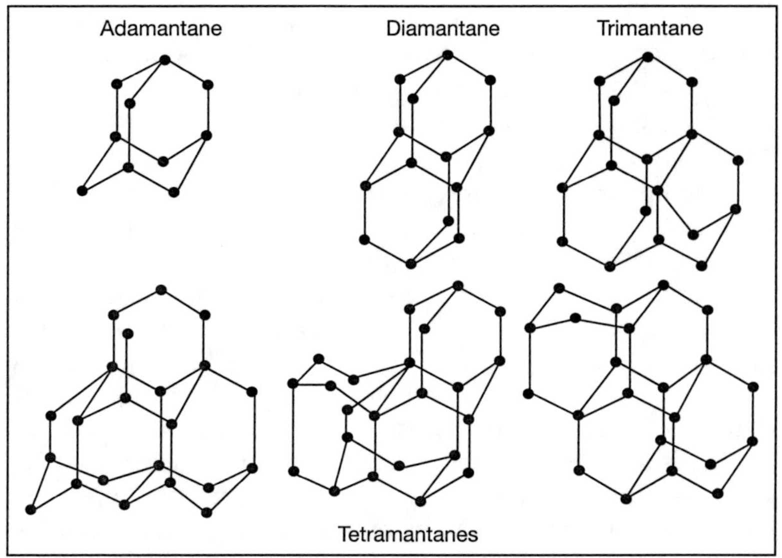

Fig. 2.11: Chemical structures of diamondoid organic nanostructures.

Applications

Diamondoids have the potential to produce different types of nanostructural shapes such as molecular-scale components of machinery like rotors, propellers, ratchets, gears, etc. Their organic nature and sublimation properties make them capable for moulding and cavity formation applications.

Higher diamondoids have different geometries and large number of attachment sites, and hence they are able to create shape derivatives. A few derivatives of diamondoids can be used as antiviral drugs. Some of the potential applications of diamondoids include drug targeting, cages for drug delivery, gene delivery and molecular machines.

2.11 Smart Nanomaterials

Smart materials are materials with one or more properties that can be modified in a controlled manner by external stimuli, such as temperature, force, moisture, pH, electric and magnetic fields. They are key materials to offer structures with smart functions and can be used as advanced functional materials.[19-21]

Smart structures show the skill to study their instantaneous environment and execute control actions in a reliable way. They find potential applications in various areas like aerospace, transportation, medical hardware, manufacturing technology and so on.

The influence of nanotechnology enhances the potential of these materials. Smart nanomaterials are materials made at the nanoscale level, which performs the functions of smart materials. These materials possess strange structures with mobile electronic charge carriers. These mobile charge carriers can move to new positions in the structure. This change can be controlled by shining light on them or applying an electric field, and the change is like a code. These smart nanomaterials can incorporate nanosensors, nanocomputers and nanomachines into their structure. As a result, they may be enabled to respond directly to their surroundings. Self-tinting automotive glass is an example of a smart material. This glass is clear most of the time, but when the sunlight reaches certain intensity level, the glass darkens to avoid the driver from being blinded. These smart materials are able to shape shift to enable a comfortable environment. For instance, smart flexible clothing for motorcyclists can turn it to hard, as it senses an impact.

There are various types of smart materials, such as piezoelectric, shape memory, thermochromic, photochromic and magnetorheological. The working of the piezoelectric smart material is based on piezoelectric effect. The shape memory materials are able to memorize their initial shape and recover it while heating. For example, if shape memory stents-tubes are attached into arteries, which expand on heating to body temperature and allow smooth blood flow. The colour of thermochromic materials changes as temperature changes, but photochromic smart materials would change colour in response to colour variations. Magnetorheological fluid materials become solid when magnetic influence is felt. Thus, they can be used in buildings or bridges to reduce the shock impact of earthquakes. Smart nanomaterials offer an array of promising applications in a variety of fields, which include healthcare, energy generation and conservation, security and terrorism defence, and implants and prostheses.

REFERENCES

1. Knauth, P. and Schoonman, J., *Nanostructured Materials: Selected Synthesis Methods, Properties and Applications* (Springer, 2002).
2. Nalwa, H.S. (Ed.), *Handbook of Nanostructured Materials and Nanotechnology* (Academic Press: San Diego, 1999).
3. Rao, C.N.R., Müller, A. and Cheetham, A.K., *The Chemistry of Nanomaterials: Synthesis, Properties and Applications* (John Wiley & Sons, 2004).
4. Rao, C.N.R., *Nanoworld*, 2010.
5. Dresselhaus, M.S. and Dresselhaus, G., *Nanostuctured Materials*, 9, 33-42 (1997).
6. Bhusion, B., *Handbook of Nanotechnology* (NY: Springer-Heidelberg, 2004).
7. Brechignac, C., Houdy, P. and Lahmani, M. (Eds.), *Nanomaterials and Nanochemistry* (Springer: NY, 2007).
8. Edelstein, A.S. and Cammarata, R.C. (ed.), *Nanomaterials: Synthesis, Properties and Applications* (IOP Publishing: Bristol and Philadelphia, 1996).
9. Lue, J.T., *Encyclopedia of Nanoscience and Nanotechnology*, Vol. X, 1-46 (2007).
10. Guozhong, Cao, *Nanostructures and Nanomaterials: Synthesis, Properties, and Applications* (Imperial College Press: London, 2004).
11. Dresselhaus, M.S., Dresselhaus, G. and Eklund, P.C., *Science of Fullerenes and Carbon Nanotubes* (Academic Press: NY, 1996).
12. Dresselhaus, M.S., Dresselhaus, G., and Saito, R., *Carbon*, 33, 883-91 (1995).
13. Iijima, S., *Nature*, 354, 56 (1991).
14. Suchita, Kalele, Goosavi, S.W., Urban, J. and Kulkarni, S.K., *Current Science,* 91(8) (2006).
15. Pradeep, T., *Nano: The Essentials* (McGraw Hill: New Delhi, 2007).
16. David, S. Goodsell, *Bionanotechnology: Lessons from Nature* (Wiley-Liss, Inc., Hoboken: New Jersey, 2004).
17. Ali, Mansoori G., Diamondoid molecules, *Adv. Chem. Phy.* 136, 207-58 (2007).
18. Mansoori, G.A., *Principles of Nanotechnology—Molecular-Based Study of Condensed Matter in Small Systems* (World Scientific Publishing: Singapore, 2005).

19. Yoshida, M. and Lahann, J., *ASC Nano*, 2(6), 1101-07 (2008).
20. John, Mongillo, *Nanotchnology*, 101 (Pentagon Press, 2007).
21. Rainer, Wasser (Ed.), *Nanoelectronics and Information Technology: Advanced Electronic Materials and Novel Devices* (WILEY-VCH Verlag GmbH & Co. KgaA: Weinheim, 2003).

3

Properties of Nanomaterials and Scaling Laws

3.1 Introduction

The main points to be discussed in this chapter are, how properties and behaviour of materials change as their characteristic dimension is reduced, especially discontinuous changes occurring at the nanoscale. A concrete knowledge about the properties of nanoparticles with particle size is essential for the understanding of the fundamental concepts of condensed matter. The discovery of new tools helps the scientists to see and arrange molecules at the nanoscale. These nanostructured materials exhibit more varied properties and behaviours than large particles of the same material.

A brief review of fundamental physics may help us to understand how nanotechnology could revolutionize diverse areas. Mainly there are two sets of theories that are significant while discussing about nanostructured materials, classical mechanics and quantum mechanics. The classical mechanics deals with the world of macroscopic things. However, quantum mechanics deals with the world of atoms and molecules. It is possible to find the exact position of an object with classical mechanics. But, quantum mechanics does not provide such information. It helps to explain the strange behaviour of atoms and molecules. Figure 3.1 shows that quantum mechanical rules dominate over classical rules as we move along the scale from large to small. The middle region between the two regimes is the realm of nanotechnology.

3.2 Size Dependence of Properties

The physical properties of nanomaterials differ considerably from that of bulk because the system size approaches quantum mechanical length scales.[1-10] Most of the size induced changes, both physical and chemical properties, are mainly due to the following reasons:

(i) Gravitational forces are negligible due to small mass of the particles, while electromagnetic forces are very strong in nanosized particles and are dominant in determining the bahaviour of particles. The other two forces, the strong nuclear force and weak nuclear force, are only significant at extremely short distances and hence become negligible in the nanoscale.

(ii) The quantum mechanical descriptions of particle motion and energy transfer is significant at nanoscale dimensions instead of the classical mechanical descriptions. The discrete energy levels of electrons, the wave like properties of matter, quantum tunnelling and uncertainty of measurement are some of the most important things that quantum mechanical models can describe.

(iii) The nanoscale objects have a very large surface area to volume ratio, so surface effects are far more significant.

(iv) At nanoscale, the influence of random molecular motion plays a much greater role than they do at the macroscale. Random molecular motion is the movement that all molecules in a substance exhibit due to their kinetic energy. At the macroscale this motion is very small compared to the size of the objects and thus is not influential in material behaviour. However at the nanoscale, these motions can be on the same scale as the size of the particles and thus have an important role on material behaviour.

Thus, the peculiar properties of nanosized objects are due to influence of random motion, high surface area to volume ratio, dominance of electromagnetic force and significance of quantum mechanics.

It is very interesting that colour of gold changes with the size of the particles. Usually gold is a shiny yellow metal and can be modified into a variety of shapes for ornaments. If we reduce the size of materials down to a critical size, their properties begin to change drastically. For instance, when the size of the gold is reduced to nano range, gold shows different colours like red, orange or purple, depending upon the particle size.

As the size varies, the particles absorb or reflect light differently, based on their energy levels. These energy levels are determined by size and bonding arrangements of particles.

The individual atoms do not have colour. The colour of a substance is determined by the wavelength of the light that reflects it, and one atom is too small to reflect light on its own. For gold, colour is based on the crystalline or atomic structure at the nanoscale, and light absorbs or reflects differently based on the thickness of the crystal. It has been observed that most of this variability begins at the nanoscale. Therefore, it is possible to control or manipulate material properties by controlling the synthesis of nanomaterials.

3.3 Properties of Nanomaterials

As mentioned earlier, matter behaves mysteriously when its size approaches the nanoscale. The peculiar properties of nanomaterials arise from many different fundamentals.[11-21] For example, the huge surface energy is responsible for the reduction of thermal stability and superparamagnetism. Increased surface scattering is responsible for the reduced electrical conductivities. Size confinement may cause a change of both electronic and optical properties of nanomaterials. The reduction of size favours an increase in perfection and, thus enhance the mechanical properties of individual nanosized materials. However, the size effects on mechanical properties of bulk nanostructured materials are far more complicated, since there are other mechanisms involved, such as grain boundary phase and stresses. An overview of the peculiar properties of nanomaterials and the fundamentals behind these properties are presented in this section.

The surface area-to-volume ratio of the material can be enhanced tremendously by simply reducing its size. The size reduction is defined in terms of top-down approach and macroscopic scaling laws are applicable in this case. An object is delineated by its boundary. Suppose a spherical object of radius r is heated by internal processes, the amount of heat will be proportional to its volume $V = (4\pi r^3/3)$ and the loss of heat to the environment will be proportional to the surface area, $A = 4\pi r^2$. If the object is divided into n small particles, the total surface area now increases to $n^{1/3} 4\pi r^2$. This is the basic reason why small mammals have a higher metabolic rate than larger ones—they need to produce more heat to compensate for its relatively greater loss through the skin in order to keep their bodies at the same steady temperature. This also explains why a few small mammals are found in the cold regions of the earth. Increase in surface area with decrease in size of materials is illustrated in example 1.

Example 1

Consider a cube of side 2 cm length having a surface area of 24 cm^2. If it is divided into 8 small cubes, its length reduces to 1 cm. But, the total surface area of these 8 cubes becomes 48 cm^2. If each of these cubes is further divided in to 10^{21} cubes, their side length reduces to 1 nm. The total surface area of 8×10^{21} cubes = $8 \times 6 \times 10^{21}$ nm^2 = 48×10^7 cm^2.

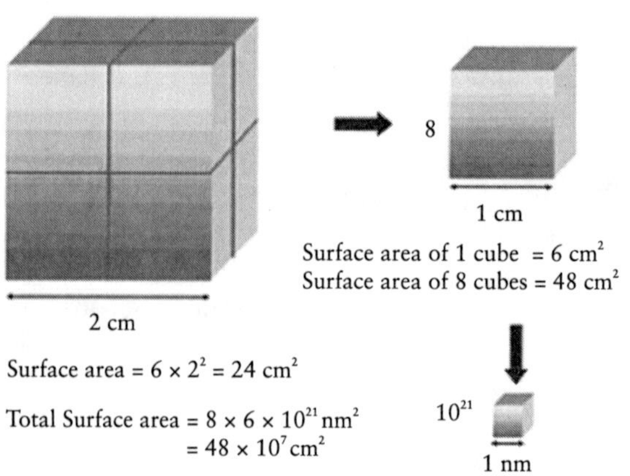

3.3.1 Chemical Reactivity

Matter is made up of atoms and the atoms situated at the surface of an object are qualitatively different from those in the bulk (figure 3.1). The surface atoms (white) have only four nearest neighbours atoms, whereas inner atoms (black) have six nearest neighbours (in the two-dimensional cross-section) of their own kind. This may have a direct impact on chemical reactivity. Usually the surface atoms are individually more reactive than their bulk neighbours, since they have some free valences (i.e., bonding possibilities). Consideration of chemical reactivity (its enhancement for a given mass, by dividing matter into nanoscale-sized pieces) suggests a discontinuous change when matter becomes 'all surface'.

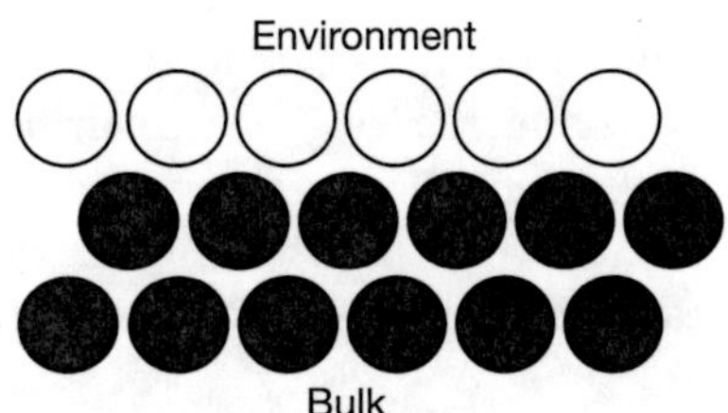

Fig. 3.1: The boundary of an object is shown as a cross-section in two dimensions.

The surface atoms easily satisfy their bonding requirements by finding reaction partners from the environment. For example, many metals become spontaneously coated with a film of their oxide by reacting with air. Hence, they are chemically more inert than pure material. The thickness of these films are typically greater than one atomic layer. A one centimetre cube of sodium taken from its protective fluid (naphtha) and thrown into a pool of water will act in a lively fashion for some time. However, if the sodium is first cut up into one micrometre cubes, most of the metallic sodium will be reacted with moist air before it reaches the water. Thus reaction time also changes at small scale. Nanosized objects exhibit very high rate of reaction because of their unique surface structure and large surface area to volume ratio.

3.3.2 Solubility

The vapour pressure P of a droplet, and the solubility of a nanoparticle, increases with diminishing radius r according to the Kelvin equation

$$k_B T \ln(P/P_0) = 2Sv/r, \quad ...(3.1)$$

where k_B is Boltzmann's constant, T the absolute temperature, P_0 the vapour pressure of the material terminated by an infinite planar surface, S the surface tension (which may itself be curvature-dependent) and v the molecular volume.

3.3.3 Melting Points

Nanoparticles of metals, semiconductors, nanowires, inert gases and molecular crystals are all found to have lower melting temperatures, when the particle size decreases below 100 nm. The lowering of melting points is due to the fact that the surface energy increases as size decreases. The transition temperature of lead titanate (Pb TiO_3) decreases as the particle size decreases as shown in figure 3.2. It shows that the bulk Curie temperature of lead titanate is retained till the particle size drops below 50 nm.

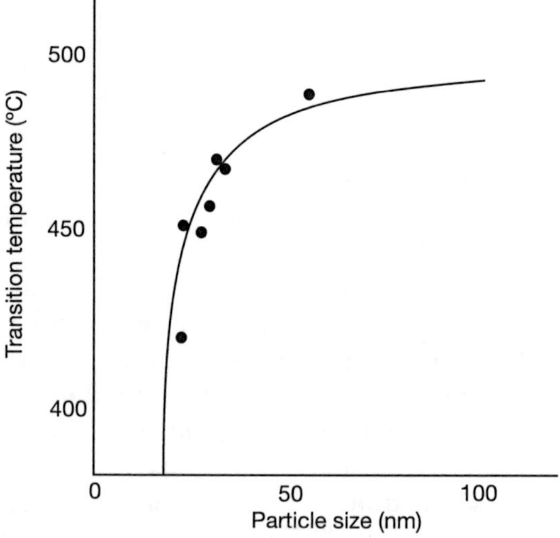

Fig. 3.2: The transition temperature as a function of particle size.

3.3.4 Electronic Energy Levels

Individual atoms have discrete energy levels and their absorption spectra correspondingly represent sharp individual lines. It is a well-known feature of condensed matter that these discrete energy levels merge into bands and the possible emergence of a forbidden energy gap (band gap) determines whether the material is a metal or a dielectric. Stacking objects with nanoscale sizes in one, two or three dimensions (nanoplates, nanofibres and nanoparticles respectively), constitute a new class of superlattices or superatoms. These are exploited in a variety of nanodevices. The superlattice gives rise to sub-bands with energies

$$E_n(k) = E_n^{(0)} + \hbar^2k^2/(2m^*) \qquad ...(3.2)$$

where $E_n^{(0)}$ is the nth energy level, k the wavenumber and m^* the effective mass of the electron, which depends on the band structure of the material. Similar phenomena occur in optics, but since the characteristic size of photonic band crystals are in the micrometre range, they are beyond the scope of nanotechnology.

3.3.5 Electrical Conductivity

The electrical conductivity of nanomaterials depends on their size. The effects of size on electrical conductivity are complex and are based on different mechanisms. These mechanisms are surface scattering including grain boundary scattering and quantized conduction. Besides, increased perfection, such as reduced impurity, structural defects and dislocations, would affect the electrical conductivity of nanomaterials.

3.3.5.1 Surface Scattering

Electrical conduction in metals or Ohmic conduction can be described by the various electron scattering. The total resistivity, ρ_T, of a metal is a combination of the contribution of individual and independent scattering, known as Matthiessen's rule:

$$\rho_T = \rho_{Th} + \rho_D \qquad ...(3.3)$$

where ρ_{Th} is the thermal resistivity and ρ_D the defect resistivity. The collision of electrons with vibrating atoms (phonons) can shift them from equilibrium lattice positions. This is the source of the thermal or phonon contribution, which increases linearly with temperature. Impurity atoms, defects and grain boundaries disturb the periodic electric potential of the lattice and may root electron scattering, which is independent of the temperature. Obviously, the defect resistivity can be further divided into impurity resistivity, lattice defect resistivity and grain boundary resistivity. Considering individual electrical resistivity directly proportional to the respective mean free path (λ) between collisions, the Matthiessen's rule can be written as:

$$1/\lambda_T = 1/\lambda_{Th} + 1/\lambda_D \quad ...(3.4)$$

Theory suggests that λ_T ranges from several tens to hundreds of nanometers. Reduction in material's dimensions would have two different effects on electrical resistivity. One is an increase in crystal perfection or reduction of defects, which would result a reduction in defect scattering and, thus, a reduction in resistivity. However, the defect scattering makes a minor contribution to the total electrical resistivity of metals at room temperature, and thus the reduction of defects has a very small influence on the electrical resistivity. The surface scattering is also contributed to the total resistivity, which has a very important role in the total electrical resistivity of nanosized materials. If the mean free electron path, λ_S, due to the surface scattering is small, then the total electrical resistivity:

$$1/\lambda_T = 1/\lambda_{Th} + 1/\lambda_D + 1/\lambda_S \quad ...(3.5)$$

In nanowires and thin films, the surface scattering of electrons results in reduction of electrical conductivity. When the critical dimension of thin films and nanowires are less than λ_S, the electron motion will be interrupted through collision with the surface. The electrons undergo either elastic or inelastic scattering. In elastic scattering, the electron does not lose its energy and its momentum or velocity along the direction parallel to the surface is preserved. As a result, the electrical conductivity remains the same as in the bulk and there is no size effect on the conductivity. The electron mean free path is terminated by impinging on the surface as the scattering is inelastic. After the collision, the path of electron is independent of the impingement direction and the successive scattering angle is random. Consequently, the scattered electron loses its velocity along the direction that is parallel to the surface or the conduction direction, and the electrical conductivity decreases. Thus, there is a size effect on electrical conduction. Moreover, an increased surface scattering would result in reduced electron mobility and, thus, an increased electrical resistivity.

It should also be noted that the surface inelastic scattering of electrons and phonons would result in a reduced thermal conductivity of nanostructures and nanomaterials, similar to the surface inelastic scattering on electrical conductivity.

3.3.5.2 Change of Electronic Structure

A reduction in characteristic dimension below a critical size, i.e. the electron de Broglie wavelength, would result in a change of electronic structure, leading to widening and discrete band gap. Such a change generally would result in a reduced electrical conductivity. Some metal nanowires become semiconducting and semiconductor nanowires become insulators, as their diameters are reduced below a certain value. Such a change can be partially attributed to the quantum size effects. For example, single crystalline Bi nanowires undergo transition from metal to semiconductor at a diameter of ~52 nm and the electrical resistance of Bi nanowires of ~40 nm decreases with decreasing temperature. GaN nanowires of 17.6 nm in diameter is found to be still semiconducting. However, Si nanowires of ~15 nm become insulating.

3.3.5.3 Quantum Transport

Ballistic conduction occurs when length of the conductor is small as compared to mean free path of electrons. In ballistic transport, each conducting channel introduces $Go = 2e^2/h = 12.9$ $k\Omega^{-1}$ to the total conductance, and also no energy is dissipated in the conduction, and there exist no elastic scattering.

If the contact resistance is greater than the resistance of nanostructures, Coulomb blockade or Coulomb charging will be occurred. Metal or semiconductor nanocrystals of a few nanometers in diameter exhibit quantum effects that give rise to discrete charging of the metal particles. Such a discrete electronic configuration permits one to pick up the electric charge one electron at a time, at specific voltage values. This Coulomb blockade behaviour, also known as "Coulombic staircase", has originated the proposal that nanoparticles with diameters below 2-3 nm may become basic components of single electron transistors (SETs). Energy is required to add a single charge to a semiconductor or metal nanoparticle, since electrons can no longer be dissolved into an infinite bulk material. For a nanoparticle surrounded by a dielectric with a dielectric constant of ε_r, the capacitance of the nanoparticle depends on its size as:

$$C(r) = 4\pi\, r\, \varepsilon_0\, \varepsilon_r, \qquad \text{...(3.6)}$$

where r is the radius of the nanoparticle and ε_0 is the permittivity of free space. The charging energy required to add a single charge to the particle is given by

$$E_c = e^2/2C(r) \qquad \text{...(3.7)}$$

Equations (3.6) and (3.7) clearly indicate that the charging energy is independent of materials.

Tunnelling conduction is another charge transport mechanism important in the nanometer range. Tunnelling involves charge transport through an insulating medium separating two conductors that are extremely closely spaced. It is because the electron wave functions from two conductors overlap inside the insulating material, when its thickness is extremely thin. But the electrical conductivity decreases exponentially with increasing thickness of insulating layer. Under such conditions, electrons are able to tunnel through the dielectric material when an electric field is applied. It is interesting that Coulomb charging and tunnelling conduction are not material properties, but they are system properties dependent on the characteristic dimensions.

3.3.5.4 Effect of Microstructure

Electrical conductivity may change due to the formation of ordered microstructure, when the size is in the nano regime. For example, polymer fibres exhibit an enhanced electrical conductivity. The enhancement is explained by the ordered arrangement of the polymer chains. Within nanometre fibres, polymers are aligned parallel to the axis of the fibres, which results in increased contribution of intramolecular conduction and reduced contribution of intermolecular conduction. Since intermolecular conduction is far smaller than intramolecular conduction, ordered arrangement of polymers with polymer chains aligned parallel to the conduction direction would result in an increased electrical conduction. A drastic increase in electrical conductivity with a decreasing diameter is found at diameters less than 500 nm. Smaller the diameter, better the alignment of polymer. A lower synthesis temperature also favours a better alignment and thus a higher electrical conductivity.

The impact of nanostructured materials on electronics is tremendous. The properties of materials can be manipulated by systematic variation of their composition, and thereby develop devices of smaller dimensions with superior functionality, improved memory density and better speed.

3.3.6 Superparamagnetism

The magnetic properties of nanoparticles differ from those of bulk materials mainly in two points. The large surface area to volume ratio offers a different local environment for the surface atoms in their magnetic coupling or interaction with neighbouring atoms, which leads to the mixed volume and surface characteristics. In nanoscale materials, several small ferromagnetic particles could possess only a single magnetic domain instead of multiple magnetic domains present in bulk ferromagnetic materials. In the case of a single particle being a single domain, the superparamagnetism arise, in which the magnetizations of the particles are randomly distributed and they are aligned only under an applied magnetic field. The alignment disappears when the applied field is withdrawn. Magnetic nanomaterials have the potential of information storage and the size of the domain determines the limit of storage density.

In certain magnetic elements, exchange interactions between the electrons of adjacent ions lead to a very large coupling between their spins. That is, above a certain temperature, the spins spontaneously align with each other. The synthesis of nanoparticles of ferromagnetic substances has led to the discovery that when the size of the particles are below a certain limit (~15 nm), the substance possesses a large magnetic susceptibility in the presence of an external field, but lacks the residual magnetism characteristic of ferromagnetism. This phenomenon is known as superparamagnetism. It is worth to note that ferromagnetic substances become unstable when the particle size reduces below the limit because the surface energy provides a sufficient energy for domains to spontaneously change polarization directions. As a result, ferromagnetics become paramagnetics and behaves differently from conventional paramagnetic. Thus, there is a lower limit to the size of the magnetic elements in nanostructured magnetic materials for data storage, typically about 20 nm. Below this limit, thermal energy due to room temperature overcomes the magnetostatic energy of the element. It will result in zero hysteresis and consequently become incapable to store magnetization-oriented information. Magnetic nanocrystals have other important applications such as in colour imaging, bioprocessing, magnetic refrigeration and ferrofluids. In magnetic refrigeration, nanocomposites moving in a magnetic field can be used instead of ozone-depleting refrigerants and energy-consuming compressors.

Mechanical heterostructured multilayers consist of alternating ferromagnetic and nonmagnetic layers (e.g., Fe-Cr, Co-Cu), which exhibit giant magnetoresistance (GMR). GMR refers to a significant change in the electrical resistance experienced by current flowing parallel to the layers when an external magnetic field H is applied. This effect occurs in a multilayer where the magnetic moments of the alternating ferromagnetic layers display an antiparallel alignment when $H = 0$. If a sufficiently strong magnetic field is applied, the magnetic moments of the ferromagnetic layers assume a parallel alignment. This change in orientation of the moments causes a change in resistance, the largest resistance occurring when the moments of the layers display an antiparallel alignment and the smallest resistance occurring when all the moments are parallel. GMR find applications in data storage and sensors.

3.3.7 Electron Confinement

One useful perspective on a definition for the appropriate length scales for nano is a regime where the chemical, physical, optical and electrical properties of matter all become size and shape dependent.[22, 23] The wave-like properties of matter become important when dealing with small things. Therefore, according to de Broglie, the wavelength associated with each object is,

$$\lambda = h/p \qquad ...(3.8)$$

where λ is the de Broglie wavelength, $h = 6.62 \times 10^{-34}$ J-s, Planck's constant and p is the momentum (mv) with m as the mass. A nanomaterial is in the "quantum confinement" regime because its size is smaller than the corresponding de Broglie wavelength of an electron or hole, or its size is less than exciton Bohr radius. Excitons are bound electron-hole pairs caused by the absorption of a photon in a semiconductor.

Quantum mechanical effects have to be considered at nanoscale dimensions. Let L_c be the critical size of the element, which compares with the wavelength (λ) of electrons, i.e., $L_c \approx \lambda = h/p$, where h and p are the Planck constant and the electron momentum respectively. In such a condition, nanoparticles (also called quantum dots) perform like large atoms and the electronic motion is restricted in three dimensions. Such a system is known as zero-dimensional system. Besides these, quantum mechanical calculations show that electron possess discrete energy levels as in an atom. But, in a solid these levels are grouped into energy bands. The cause of new electronic properties may be explained on the basis of quantum mechanics with Heisenberg's uncertainty relation. The position and momentum of an electron cannot be determined simultaneously with accuracy. So the average energy is not determined by its chemical origin but only by the dimension. Electrons localized as "particle in a box" with zero dimensional quantum dots lose their freedom in all three dimensions, leading to discrete energy states, as long as their energy is not great enough to overcome this confinement. Whenever the size of the object is equal in magnitude of the electron wavelength or even smaller, quantum effects govern the wave propagation of the system. This effect is called the Quantum Size Effect (QSE).

Bohr Radius

The equation for the Bohr radius of an electron moving in a condensed phase is given by

$$r_B = 4\pi\varepsilon_0\varepsilon\hbar^2/mq^2 \quad ...(3.9)$$

where $\varepsilon_0 = 8.85 \times 10^{-12}$ F/m (permittivity), $\hbar = 1.054 \times 10^{-34}$ J·s (Planck's constant over 2π), $m_e = 9.11 \times 10^{-31}$ kg (mass of a free electron) and $q = 1.602 \times 10^{-19}$ C.

On substituting all the above values,

$$r_B = 5.28 \times 10^{-11} \text{ metres}$$

$$= 0.528 \text{ Angstroms} \quad ...(3.10)$$

This is the standard Bohr radius.

But for the case of an exciton (electron hole pair) in a semiconductor, just replace the mass of the electron with the effective mass of the exciton.

$$1/m_{eff} = [(1/m_e) + (1/m_h)],$$

where m_e and m_h are the effective masses of electron and hole in the material.

Example 2

Calculate the Exciton Bohr radius for GaAs. Given that, $m_e = 0.067\ m_0$, $m_h = 0.45\ m_0$ and $\varepsilon = 12.4$ (literature values).

Solution:

Let $1/m_{eff} = [(1/m_e) + (1/m_h)]$

$$= (1/0.067m_0) + (1/0.45\ m_0)$$

Therefore the effective mass $m_{eff} = 0.058\ m_0$

The Exciton Bohr radius for GaAs

$$r_B = (4\pi\varepsilon_0\varepsilon\hbar^2/0.058\ m_0q^2)$$

$$= [4\pi(8.85 \times 10^{-12})(12.4)(1.054 \times 10^{-34})^2]/$$

$$[(0.058)(9.11 \times 10^{-31})(1.602 \times 10^{-19})^2]$$

$$= 11.3 \text{ nm}$$

Example 3

Find the Exciton Bohr radius for CdSe. Using literature values, $m_e = 0.13\ m_0$, $m_h = 0.45\ m_0$ and $\varepsilon = 9.4$.

Solution:

Let $1/m_{eff} = 1/m_e + 1/m_h$

$= (1/0.13\ m_0) + (1/0.45\ m_0)$

Therefore the effective mass $m_{eff} = 0.1\ m_0$

The Exciton Bohr radius for CdSe

$$r_B = (4\pi\varepsilon_0\varepsilon\hbar^2/0.058\ m_0q^2)$$

$$= [4\pi\ (8.85 \times 10^{-12})(9.4)(1.054 \times 10^{-34})^2]/[(0.1)(9.11 \times 10^{-31})(1.602 \times 10^{-19})^2]$$

$$= 4.97\ \text{nm}$$

Because of the variable relative dielectric constant and effective masses of carriers in a semiconductor, there will be various exciton Bohr radii. Consider the case of GaAs, the bulk exciton Bohr radius is $r_B = 11.3$ nm. The lattice constant of GaAs which crystallizes in the zinc blende form, a = 5.65 Å. Thus the radius of the electron orbit is about 20 unit cells away from the counterpart hole. That is, in semiconductors the electrons can move very far away from its counterpart due to the screening effect of the material.

Table 3.1: Bohr radius of the excitons (r_B) in semiconductors

Sl. No.	Material	Band gap (eV)	r_B (nm)
1.	Si	1.17	4.3
2.	GaAs	1.518	11.3
3.	CdSe	1.84	4.9
4.	CdS	2.583	2.8
5.	ZnSe	2.82	3.8
6.	AgBr	2.684	4.2
7.	CuBr	3.077	1.2
8.	CuCl	3.395	0.7

The typical values of r_B range from a few to a few hundred nanometres. Bohr radius of the excitons (r_B) in semiconductors are listed in Table 3.1. Therefore, it is practically possible to create particles whose radius r is smaller than the Bohr radius. The de Broglie wavelength or exciton Bohr radius is natural length scales which helps to compare the physical size of a nanomaterial. In general, objects with dimensions smaller than these natural length scales will exhibit quantum confinement effects. Excitons can be seen in the absorption spectrum of bulk semiconductors. They generally appear just below the band edge of the semiconductor. This is because the energy of the exciton is lower than the band edge transition by its binding energy. That is, the exciton energy $E_{exc} = E_g - E_{bind}$, where E_g is the bulk band gap of the semiconductor and E_{bind} the exciton binding energy.

The quantum confinement has applications in semiconductors, optoelectronics and non-linear optics. The spherical-like shape of nanocrystals produces surface stress, resulting in lattice relaxation (expansion or contraction) and change in lattice constant. It is known that the electron energy band structure and band gap are sensitive to lattice constant. The lattice relaxation caused by nanocrystal size can affect its electronic properties.

3.3.8 Integrated Optics

Light can be confined to a channel or a plate made from a transparent material having a higher refractive index than that of its environment. Effectively, light propagates in such a structure by successive total internal reflections at the boundaries. The diameter of channel (fibre) or the plate thickness is less than the wavelength of the light. Below a certain minimum diameter or thickness (the cut-off), however, typically around one third of the wavelength of the light, propagation is no longer possible. The science and technology of light guided in thin structures is called integrated optics and fibre optics, and sometimes nanophotonics. However, the cut-off length is several hundred nanometres, and does not fall into the nano region as it is currently defined.

3.3.9 Optical Properties

The reduction of size of materials has found effects on their optical properties. The change in optical properties is mainly due to the surface plasmon resonance and the increased energy level spacing. The coherent excitation of entire free electrons in the conduction band may produce an in-phase oscillation, called surface plasmon resonance. When the size of a metal nanocrystal is smaller than the wavelength of incident radiation, a surface plasmon resonance is generated. Thus, plasmon resonance depends on the particle size.

3.3.9.1 Surface Plasmon Resonance (SPR)

When materials absorb light of resonant wavelength, the free electrons in the conduction band starts to vibrate and dissipate energy. This process usually occurs at the surface of the material, and hence the name surface plasmon resonance. The oscillations of the free electron cloud are named as plasmons. The electric field of the absorbing light causes polarization of the conduction electrons and a net charge difference occurs at the surface of nanoparticles, which in turn acts as a restoring force. As a result, a dipolar oscillation of electrons is generated with a specific frequency. The surface plasmon resonance is a dipolar excitation of the entire particle between the negatively charged free electrons and its positively charged lattice. The energy of the surface plasmon resonance depends on both the free electron density and the dielectric medium surrounding the nanoparticle. Schematic representation of plasmon resonance absorption is shown in figure 3.3.

Usually nanoparticles exhibit SPR in the visible part of electromagnetic radiation. This shows that a certain wavelength of visible light is absorbed by nanoparticles and is converted into surface plasmons. Small nanoparticles absorb blue-green wavelength of light, but they reflect red light. When size of the nanoparticles increases, wavelength of plasmon resonance absorption moves to longer wavelength (red wavelength) side. Now red light is absorbed and blue light is reflected, resulting in pale blue or purple colour for the particles. When particle size increase towards critical limit, SPR wavelength shifts to the IR spectrum of the radiation and visible lights are reflected. These properties find potential applications in biosensors.

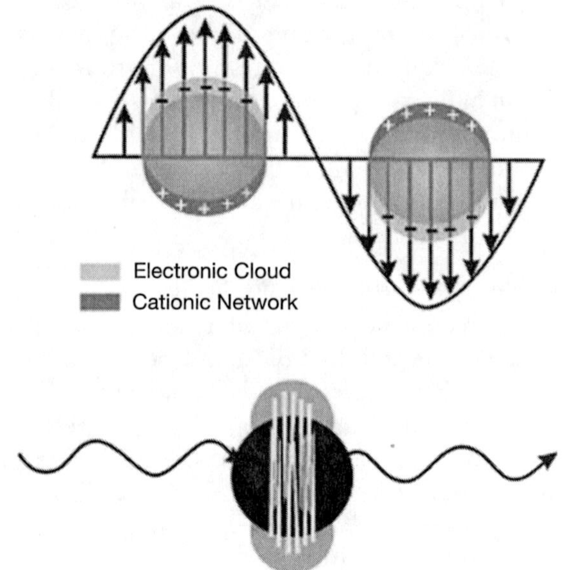

Fig. 3.3: Plasmon resonance absorption.

Source: www.willets.cm.utexas.edu

It is also reported that metal nanowires have surface plasmon resonance properties, similar to nanoparticles. However, metal nanorods exhibit two SPR modes, one due to transverse and the other due to longitudinal excitations. The wavelength of transverse mode is usually set at nearly 520 nm for gold and 410 nm for silver. However, their longitudinal modes can be tuned to span from visible region to near IR region by controlling their aspect ratios.

3.3.9.2 Quantum Size Effects

Unique optical properties of nanomaterials may also arise from quantum size effect. As discussed in section 3.3.7, when the size of a nanocrystal (i.e. a single crystal nanoparticle) is smaller than the de Broglie wavelength, electrons and holes are spatially confined and electric dipoles are formed, and discrete electronic energy level would be formed in all materials. The energy gap between adjacent energy levels increases as size decreases, similar to a particle in a box.

In semiconductor nanoparticles, the quantum size effect is predominant and the band gap increases with a decreasing size, and the interband transition shifts to higher frequencies. In a semiconductor, the forbidden energy gap is of the order of a few electron volts and it increases quickly with a decreasing size. It is found that the optical absorption edge and the luminescence peak position shift to a higher energy as the particle size reduces. Such a size dependence of absorption peak has been widely used in determining the size of nanocrystals. Figure 3.4 shows that the band gap of silicon nanowires increases as diameter of the nanowire reduces.

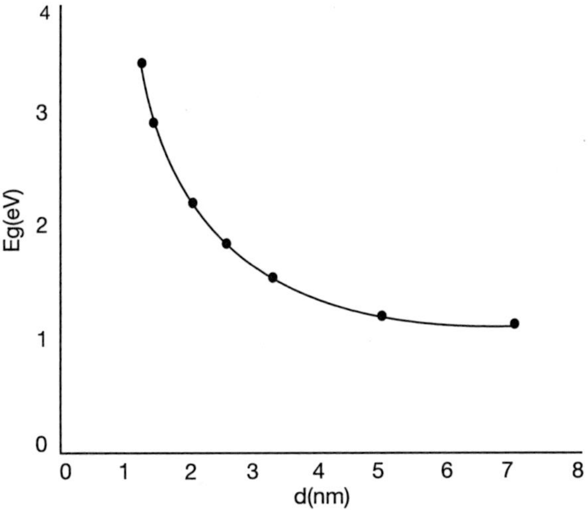

Fig. 3.4: The band gap of silicon nanowires as a function of the diameter.

The quantum size effect is also observed in metal nanoparticles. However, the localization of the energy levels can be observed only at very small size, a few nanometre. The conduction band of a metal is half filled and the density of energy levels is high. Hence, it is hard to observe gap in energy levels within the conduction band (intraband transition). If the size of metal nanoparticle is made small enough, the continuous density of electronic states is broken up into discrete energy levels. The spacing (δ) between energy levels depends on the Fermi energy of the metal (E_F) and on the number of electrons in the metal (N) as given by:

$$\delta = (4E_F/3N) \qquad ...(3.11)$$

where E_F is around 5 eV in most of the metals. The discrete electronic energy level in metal nanoparticles has been observed in far-infrared absorption measurements of gold nanoparticle. If the diameter of nanowires or nanorods reduces below the de Broglie wavelength, quantum size effect can play a vital role in determining the energy level just as for nanocrystals. For example, the absorption edge of Si nanowires has a significant blue shift with sharp, discrete features and silicon nanowires also have shown relatively strong band-edge photoluminescence. Besides the size confinement, light emitted by the nanowires is polarized along their direction.

3.3.10 Mechanical Properties

The mechanical properties of a solid depend on the density of dislocations, interface-to-volume ratio and grain size. A decrease in grain size significantly affects the yield strength and hardness. Many of the mechanical properties of materials, such as hardness, elastic modulus and fatigue strength are modified at nanoscale dimensions. It is worth mentioning that mechanical strength of nanowires approaches to the theoretical value when the diameter is less than 10 microns. The enhancement of mechanical strength of nanowires or nanorods is mainly due to two mechanisms. Firstly, the increase of strength is due to the high internal perfection of the nanowires. Thermodynamically, imperfections in crystals are highly energetic and should be eliminated from the perfect crystal structures. It is possible to eliminate such imperfections if the size is small. Another mechanism for better mechanical strength is the perfection of the side faces of nanowires. In general, smaller structures have less surface defects. It is particularly true when the materials are made through a bottom-up approach.

Figure 3.5 shows that Young's modulus decreases with porosity for nanocrystalline Cu. Also, there is a change in the yield stress and tensile ductility. Usually reduction in grain size cause an increase in ductility, however, it reduces for most grain sizes < 25 nm for metals. This reduction in ductility for nanocrystalline materials is due to artifacts from processing (e.g., pores), tensile instability and crack nucleation. It is hard to synthesize nanomaterials that are free from the artifacts, which mask the inherent mechanical properties.

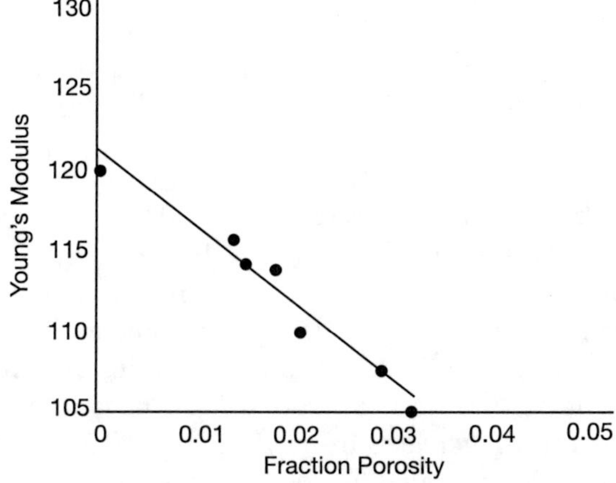

Fig. 3.5: Young's modulus as a function of porosity for nanocrystalline Cu.

Many of the nanomaterials exhibit superhardness. A number of superhard nanocomposites can be synthesized from nitrides, borides and carbides by plasma-induced chemical vapour deposition or physical vapour deposition. These superhard nanocomposites can be used in hard protective coatings. Carbon nanotubes exhibit excellent mechanical properties such as high Young's modulus and high tensile strength. Nanocomposites made from carbon nanotubes will possess amazing mechanical properties, such as high Young's modulus, stiffness and flexibility. A detailed discussion of properties of carbon nanotubes is given in section 6.4.1. Polymer nanocomposites made from nanosize fillers can produce drastic improvements in mechanical properties like yield stress and Young's modulus.

One of the significant applications of nanomaterials is in superplasticity, the capability of a polycrystalline material to undergo extensive tensible deformation without fracture. Nanomaterials possess different superplasticity from that of large-grained bulk materials. For example, near perfect superplasticity was observed in pure nanocrystalline copper, prepared by powder metallurgy. Twinning is observed in nanosized aluminium grains, which have never been found in large size particles.

3.3.11 Thermodynamic Properties

Thermodynamic properties of nanocrystals depend on its surface-to-volume ratio. The reduced coordination number of the surface atoms increases the surface energy. Hence atom diffusion occurs at relatively lower temperatures. The melting point of gold particles lowers to about 300°C for particles with diameters of less than 5 nm, much lower than the normal melting point of 1063°C for gold. Specific heat, entropy and thermal expansion of nanoparticles are much different from their bulk form. Nano crystalline palladium (6 nm) shows enhancement of specific heat from 29 per cent to 53 per cent when temperature increases from 150 K to 300 K. For the same temperature variation, nano crystalline copper shows enhancement of specific heat from 9 per cent to 11 per cent. Since copper and palladium are dia and paramagnetic substances, the enhanced specific heat may be from vibrational or configurational entropy effects.

3.4 Scaling Laws

The concepts of scaling laws are crucial in nanotechnology. These are observations of physical parameters which changes considerably depending on the scale (size) being considered.[24-31] Most physical properties are influenced by the scale, although some of the properties are retained irrespective of the scale. Scaling laws are significant while designing a very large or small construct and proper care is needed to extend principles of one construct to another. The study of scaling laws in nanotechnology is significant because it helps to exploit the extraordinary properties and behaviours of nanomaterials.

Scaling laws are simple observations of various modifications of physical variables at different sizes. For example, a flea can jump many times of its height, but an elephant can't jump that much because smaller things are less affected by gravitational force. Scaling laws can provide a simple way to know about the nanoscale, while engineering requires more intricate calculations. The physical magnitudes of nanoscale systems are extremely different from those of macroscale systems. Some of these magnitudes can be estimated by applying scaling laws to the values for macroscale systems. Moreover it is possible to find the magnitude of nanoscale systems by applying scaling laws to the values of macroscale systems.

When objects change from the macroscopic to the microscopic scale, the ratio of forces, strength, speed, etc. will change. The reason is that some of the fundamental parameters changes as dimensions are reduced. Suppose force is like a piston supplied with pressure, the transmitted force is proportional to the area of the piston. Suppose the stress transmitted through the body is kept constant as the size of the body is reduced. Thus, force = SL^2 (S is the applied stress, assumed constant). Hence the transmitted force is proportional to area or L^2, where L is some characteristic length dimension. Thus, forces for a constant stress or pressure get smaller as the length scale is reduced. For example, a stress of 10^{10} N/m^2 scales to 10^{-8} N/nm^2, i.e. 10 nN/nm^2.

Scaling laws which are applicable to mechanics, fluids, electromagnetism and thermodynamics are discussed in the following sections.

3.4.1 Mechanics

Let L be the linear dimension of an element, and assume that variation of the linear dimensions is proportional to L, so that area A is proportional to L^2,

$$A \sim L^2 \qquad ...(3.12)$$

Also, volume V is proportional to L^3,

$$V \sim L^3 \qquad ...(3.13)$$

Mass of an object m = ρV, where ρ is the density of the material. Hence, the mass m is proportional to L^3

$$m \sim L^3 \qquad ...(3.14)$$

Let us consider the variation of forces with L. The gravitational force, one of the forces in nature, is given by $F_g = mg$, where g is the acceleration due to gravity.

$$F_g \sim L^3 \qquad ...(3.15)$$

The pressure exerted by a body on the surface of the earth $P_g = F_g/A$, then

$$P_g \sim L^3/L^2 = L \qquad ...(3.16)$$

Adhesion forces are mainly due to the presence of van der Waals type forces. These forces dominate over gravitational forces at nanoscale dimensions because gravitational forces are negligible compared to molecular forces. Let us consider two infinite flat slabs separated by a distance x, the force of attraction experienced $F_{vdw} = H/(6\pi x^3)$, where H the Hamaker constant depends on nature of the medium between the slabs. The value of H is 10^{-19} J in air and 10^{-20} J in water. This equation is applicable for small values of x (2 to 10 nm). It is evident that the force of attraction F_{vdw} varies with the contact area.

$$F_{vdw} \sim L^2 \qquad ...(3.17)$$

Equations (3.15) and (3.17) shows that forces F_g and F_{vdw} act differently with L, and hence their values also vary with L.

$$F_{vdw}/F_g \sim L^{-1} \qquad ...(3.18)$$

It is clear that adhesion force dominates over the gravitational force at low values of L, and hence gravitational forces may be neglected at nano dimensions. The adhesion force and gravitational force are equal at a critical value, which depends on thickness as well as nature of the medium between the solids.

When one surface is sliding over other, the force of friction at the macroscopic scale, $F_{fr} = \mu F_g = \mu mg$, where μ is a constant, called the coefficient of friction.

$$F_{fr} \sim L^3 \qquad ...(3.19)$$

So F_{fr} is independent of the area of contact.

The spring constant k of a nanocantilever varies with its characteristic linear dimension as l, and its mass m as l^3. Hence the resonant frequency of its vibration $\omega_0 = \sqrt{k/m}$ varies as 1/l.

This ensures a fast response—in effect, nanomechanical devices are extremely stiff. The figure of merit (quality factor) Q can be many orders of magnitude greater than the values encountered in conventional devices. On the other hand, under typical terrestrial operating conditions water vapour and other impurities may condense on moving parts, increasing drag due to capillary effects, and generally degrading performance.

The kinetic energy of a body moving with a velocity v is given by $E_k = mv^2/2$.

At constant velocity, $E_k \sim L^3$ ...(3.20)

When $v \sim L$, E_k varies as

$E_k \sim L^5$ ...(3.21)

The gravitational potential energy of a body at a height h is given by $E_p = mgh$.

If height h is constant, $E_p \sim L^3$ ...(3.22)

When $h \sim L$, E_p varies as

$E_p \sim L^4$ ...(3.23)

The potential energy of a spring is given by $E_p = k\delta L^2$, and

$E_p \sim L^2$ ...(3.24)

If the moment of inertia of a rotating body is $I \sim mL^2$, then,

$I \sim L^5$ ...(3.25)

Kinetic energy of rotation $K = (1/2)I\omega^2$, where ω is the angular velocity, then

$K \sim L^5$ ...(3.26)

This equation shows that the rotational kinetic energy of small systems decreases extremely fast with size at constant angular velocity. On the other hand, in a rotating system, the kinetic energy stored in small systems is very less as compared to large systems.

Consider the lowest eigen frequency (ν) state, where the length of the device equals λ/4 or λ/2. The phase velocity of the wave $V = \lambda\nu$, so that

$\nu \sim L^{-1}$ ...(3.27)

This relation shows that resonance frequencies are large in small systems.

Consider a cantilever beam, which is loaded by its own weight. The deflection length δ of the cantilever beam varies as in eq. (3.17).

$$\delta \sim L^2 \quad ...(3.28)$$

This shows that bending of small system will be several times less due to its own weight.

3.4.2 Fluids

It is interesting that all bodies move in a fluid, except in vacuum, and the fluid parameters change with L. If a body falls freely through a viscous fluid, after a some time, it will attain a constant velocity called terminal velocity v_t. The transient time (τ) is proportional to v_t. The terminal velocity of a spherical body of radius r is given by $v_t = 2\rho g r^2/9\eta$, where η is the coefficient of viscosity of the medium and ρ is the density. Then,

$$v_t \sim L^2 \quad ...(3.29)$$

$$\tau \sim L^2 \quad ...(3.30)$$

Thus, the viscous forces quickly damp any motion of very small objects. Tiny objects remain stationary in air or they move along with air flow. But at high velocities, transition occurs from streamline flow to turbulent flow, given by the Reynold number R_e.

$R_e = \rho v_c d/\eta$, where ρ is the density of the fluid, v_c is the critical velocity, d is the diameter of the tube in which the fluid is made to flow and η is the co-efficient of viscosity of the fluid. If $v_c \sim L$, then

$$R_e \sim L^2 \quad ...(3.31)$$

When fluids flow through pipes, the transition occurs from streamline to turbulent flow when R_e is 10^3 or above. The turbulence vanishes in small systems in which liquids flow.

Let us consider a particle that moves a distance L by diffusion, then the diffusion time $\tau_{diff} = L^2/\alpha D$, where α nd D are geometrical and diffusion constants respectively. Then,

$$\tau_{diff} \sim L^2 \quad ...(3.32)$$

This scaling is applicable for particle and thermal diffusion.

3.4.3 Electromagnetism

Electrical resistance of a conductor is given by R = ρL/A, where ρ is the resistivity of the material of the conductor, L is the length and A is the area of cross section of the conductor. Thus

$$R \sim L^{-1} \quad ...(3.33)$$

When a potential difference V is applied across a conductor, an electric current I flows through it. Then by Ohm's law, I = V/R, then

$$I \sim L \quad ...(3.34)$$

By Joule's law, the power dissipated by the conductor $W = I^2Rt$,

$$W \sim L \quad ...(3.35)$$

The electrical power dissipated per unit area varies as

$$W_{un} \sim L^{-1} \quad ...(3.36)$$

This equation shows that power dissipated per unit area is large for small systems. It is interesting to note that this effect is crucial in microelectronics, where the reduction of size of the components cause more power dissipation. It is possible to minimize this effect by reducing the applied voltage.

The electric field E = V/L varies as (when V remains constant)

$$E \sim L^{-1} \quad ...(3.37)$$

The capacitance of a parallel plate capacitor is given by $C = \varepsilon_0 A/x$, where A is the area of the plates and x is the space between them. Then capacitance varies like

$$C \sim L \quad ...(3.38)$$

When a potential difference V is applied across the capacitor, it charges. The charge on each plate Q = CV varies like

$$Q \sim L \quad ...(3.39)$$

Then, the energy stored in the capacitor is given by $E = Q^2/2C$, and

$$E \sim L \quad ...(3.40)$$

This shows that the energy storing capacity decreases with the size of the capacitor. Consider capacitors having constant charge density (Q/A), so that $Q \sim L^2$ and hence E varies as

$$E \sim L^3 \quad ...(3.41)$$

The electric force between the capacitor plates is given by $(\partial E / \partial x)$, then F varies like

$$F \sim L^2 \quad ...(3.42)$$

The magnetic field in a solenoid is given by $B = \mu In/L$, where n is the number of turns of wire, L is the length of the solenoid and I is the current through the solenoid. If the solenoid shrinks, n may remain constant. But the cross section of the coil decreases and hence, I varies. Let us assume that the current density (I/A) is constant. So that $I \sim L^2$, and the magnetic field varies like

$$B \sim L \quad ...(3.43)$$

The magnetic energy stored in the solenoid is given by $E_{mag} = B^2V/2\mu$, where V is the volume of the magnetic field. Since $V \sim L^3$, magnetic energy varies like

$$E_{mag} \sim L^5 \quad ...(3.44)$$

$$F_{mag} \sim L^4 \quad ...(3.45)$$

When going to small dimensions, the number of turns of wires depends on L and it is hard to manufacture very thin wires. Moreover, the dissipation of energy is large at small dimensions.

3.4.4 Thermodynamics

The energy E required to rise the temperature T of a system is proportional to the mass, then

$$E \sim L^3 \quad ...(3.46)$$

The power dissipation P due to conduction or radiation is proportional to the area, so that

$$P \sim L^2 \quad ...(3.47)$$

Thermal losses depend on temperature differences between the system and its surroundings. The time τ required to normalize the temperature is proportional to the square of its linear dimensions, i.e.

$$\tau \sim L^2 \quad ...(3.48)$$

As discussed earlier, systems at nanoscale dimensions offer large surface area to volume ratio, and hence the surface influences both physical and chemical properties of the system. Besides these, the melting point decreases as linear dimension decreases. For a spherical system, the melting point T_m, reduces like

$$T_m \sim 1 - f/L, \qquad ...(3.49)$$

where f is a constant, its value is about 1 nm for 3d and 5d elements.

3.5 Device Performance

It is interesting that analysis of device performance starts by studying how key parameters scale with device length: area (power and thermal losses) as length squared, volume and mass as length cubed, electromagnetic force as length to the fourth power, natural frequency as inverse length, and so on. All these relationships are used to derive the way a device's performance scales as it is designed smaller. When objects become very small, the number of entities conveying suitable information also becomes small. Small signals are more vulnerable to noise. Repetition of a message is the simplest way of overcoming noise.

Example 4

Consider a restrained 1.0 cm cube of a material weighing 1.0 gm. We can push it by a force of 1.0 Newton. Suppose same stress is applied to a 1.0 nm cube of the same material. What force do you apply?

Solution:

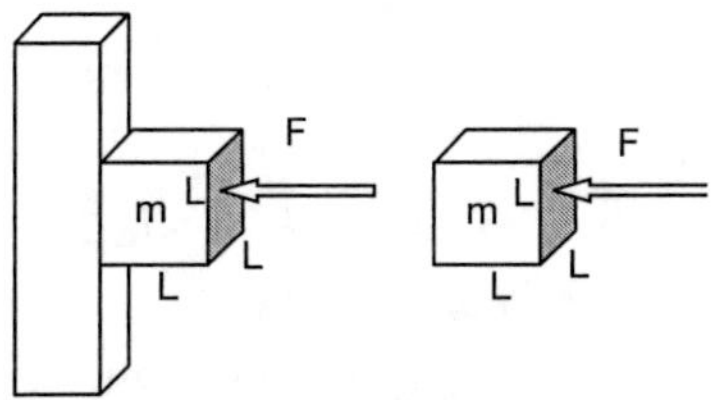

In the macroscopic scale, the stress is $1.0/(0.01 \times 0.01)$ N/m^2 $= 1.0 \times 10^4$ N/m^2. But in the case of nanometre cube, the force required to produce the same stress = stress × area = $1.0 \times 10^4 \times (10^{-9})^2 = 10^{-14}$ N. This example shows that our assumption of a constant stress is important.

Example 5

Suppose a pendulum weight is suspended from a cable. If the linear dimensions of everything is doubled, what would be the new stress?

Solution:

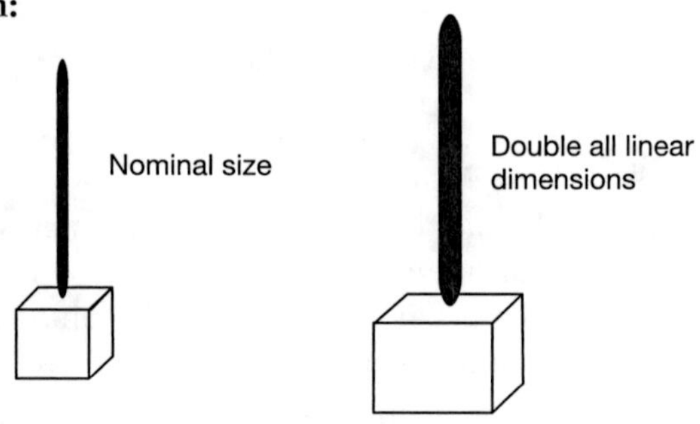

Wt = $L^3\rho$	Wt = $8L^3\rho$
Rope area = $\pi D^2/4$	Rope area = $\pi D^2 = \pi(2D^2)^2/4 = \pi D^2$
Stress $S_1 = 4L^3\rho/\pi D^2$	Stress $S_2 = 8L^3\rho/\pi D^2 = 2S_1$

Consider the original cube of side L and the cable of diameter D. The stress in the cable is $4L^3\rho/\pi D^2$. When all the linear dimensions are doubled, the new stress will be $S_2=8L^3\rho/\pi D^2= 2S_1$. Thus, a system with all dimensions doubled would also double the stress in the support cable.

REFERENCES

1. Adamson, A.W., *Physical Chemistry of Surfaces* (Wiley: NY, 1976).
2. Alhassid, Y., *Rev. Mod. Phys.*, 72 895 (2000).
3. Autumn, K., *MRS Bull.*, 32 473 (2007).
4. Baumberger, T. and Caroli, C., *MRS Bull.*, 23 41 (1998).
5. Drexler, K.E., *Phys. Educ.*, 40 339 (2005).
6. Drexler, K.E., *Nanosystems* (John Wiley and Sons, NY, 1992).
7. Kendall, K., *Science*, 263, 1720 (1994).
8. Kuno, M., *Introduction to Nanoscience and Nanotechnology: A Workbook*, 2005 (www.e-booksdirectory.com).
9. Melissinos, A.C., *Physics of Mordern Technology* (Cambridge: Cambridge University Press, 1990).

10. Ramsden, J., *Nanotechnology*, Ramsden & Ventus Publishing ApS (2009).
11. Howland, R.S. and Kirk, M.D., in *Encyclopedia of Materials Characterization*, eds. C.R. Brundle, C.A. Evans Jr., and S. Wilson, Butterworth-Heinemann, Stoneham, M.A., p. 85 (1992).
12. Lang, H.P., Hegner, M., Meyer, E., and Gerber,Ch., *Nanotechnology* 13, R29 (2002).
13. Herring, C., *Structure and Properties of Solid Surfaces*, University of Chicago, Chicago, IL, p. 24 (1952).
14. Fougere, G.E., Weertman, J.R., and Siegel, R.W., *NanoStructured Mater*: 3, 379 (1993).
15. Weertman, J.R., Niedzielka, M., and Youngdhl, C., *Mechanical Properties and Deformation Behavior of Materials Having Ultra-Fine Microstructures*, Kluwer, Boston, M.A., p. 241 (1993).
16. Hahn, H., Mondal, P., and Padmanabhan, K.A., *NanoStructured Muter*: 9, 603 (1997).
17. Konstantinidis, D.A. and Aifantis, E.C., *NanoStructured Muter*: 10, 11, 11 (1998).
18. Kreibeg, U. and Vollmer, M., *Optical Properties of Metal Clusters*, Vol. 25 (Springer-Verlag, Berlin, 1995).
19. Ferry, D.K., Grubin, H.L., Jacoboni, C.L., and Jauho, A.P., (eds.), *Quantum Transportin Ultrasmall Devices* (Plenum Press, New York, 1994).
20. Guozhong, Cao, *Nanostructures and nanomaterials* (Imperial College Press, 2004).
21. Bhusion, B., *Handbook of Nanotechnology* (Springer-Heidelberg: NY, 2004).
22. Sze, S.M., *Physics of Semiconductor Devices* (Wiley: NY, 1969).
23. Wichmann, E.H., *Quantum Physics* (McGraw Hill: NY 1971).
24. Zhong, Lin Wang, *Characterisation of Nanophase Materials*, eds. (Wiley–VCH Verlag GmbH, 2000).
25. Wautelet, M., *Eur. Phys. J. Appl. Phys.*, 29 51 (2005).
26. Wautelet, M., *Eur. J. Phys.*, 16 283 (1995).
27. Wautelet, M., *Eur. J. Phys.*, 20 L29 (1999).
28. Wautelet, M., *Eur. J. Phys.*, 22 601 (2001).
29. Wautelet, M., *Eur. J. Phys.*, 29 467 (2008).
30. Wautelet, M., *Les Nanotechnologies* (Paris: Duno, 2007)
30. Wautelet, M. and Duvivier, D., *Eur. J. Phys.*, 28, 953 (2007).
31. Hierold, C., Micromech, J., *Microengg*, 14, S1-S11 (2004).

4

Synthesis of Nanomaterials

4.1 Introduction

The ability to fabricate nanostructures with high precision is significant to the advancement of nanoscience and nanotechnology. The investigation of new properties of nanostructures and their applications are possible only when materials are synthesized with suitable size, morphology and chemical composition. In this chapter, different techniques for the synthesis of nanomaterials are presented along with a brief procedure.

Mainly, there are two approaches for the fabrication of nanomaterials, top-down and bottom-up techniques. In top-down process, usually a bulk material is taken and machined it to modify into the desired size, shape and product. Ball milling is an example of top-down technique, where macrocrystalline structures are broken down to nanocrystalline structures without changing the original integrity of the material. Lithography may be considered as a hybrid approach because growth is bottom-up while etching is top-down. But nanolithography is commonly a bottom-up approach. Both approaches play a very important role in nanotechnology. There are advantages and disadvantages in both approaches. The imperfection of the surface structure is the drawback of top-down approach. For instance, in lithographic techniques, the processed patterns contain crystallographic damages in addition to the defects, which occur during the etching process. Such imperfections would have a significant impact on physical and chemical properties of nanomaterials, since their surface to volume ratio is very large.

The bottom-up approach is represented by self-assembly or molecular assemblers. In this approach, physical and chemical forces working at the nanoscale enable the basic units to assemble into larger structures. In crystal growth, atoms, ions and molecules after impinging onto the growth surface assemble into crystal structure one after another. Bottom-up approach promises a better chance to obtain nanostructures with less defects and imperfections. The nanostructures made from this process are in a state of thermodynamic equilibrium due to the reduction of Gibbs free energy. The bottom-up approach is possible today using various techniques, such as epitaxial growth, laser ablation, physical and chemical vapour deposition and sol-gel technology.

4.2 Lithography

Lithography is a process for patterning various layers, such as conductors, semiconductors or dielectrics, on a surface. Nanopatterning expands traditional lithographic techniques into the submicron scale. A typical integrated circuit consists of various patterned thin films of metals, dielectrics and semiconductors on various substrates such as silicon, gallium arsenide or germanium. In lithography, radiation sensitive polymer materials (resists) are used to make circuit patterns on the substrates. Schematic representation of the lithographic process sequence is shown in figure 4.1.

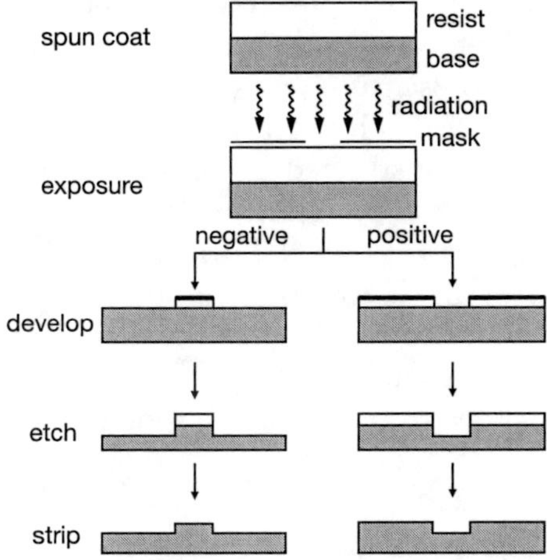

Fig. 4.1: Schematic representation of the lithographic process.[1]

The resist material is applied as a thin coating over the substrate (wafer) and then heated to remove the casting solvent. The resist film formed on the wafer is then exposed through a mask (photo- and X-ray lithography) or directly with finely focused electron beams. This film is subsequently developed using a developer solvent to form 3D images. The exposure may render the resist film more soluble in the developer, thereby producing a positive-tone image of the mask. Conversely, it may become less soluble upon exposure, resulting in the generation of a negative-tone image. When the resist image is transferred to the substrate (by etching and related processes), the resist film that remains after the development, act as a protective mask. The resist film must "resist" the etchant and protect the underlying substrate while the bared areas are being etched. The remaining resist film is finally stripped, leaving an image of the desired circuit in the substrate. The process is repeated many times to fabricate complex semiconductor devices. Photolithographic technology is the most commercially advanced form of nanolithography. Many other fabrication techniques, such as electron beam lithography, nanoimprint lithography and dip-pen technology, are also used.

4.2.1 Electron Beam Lithography

Electron beam lithography technique employs an electron beam to scan over the substrate surface, which is coated with a thin layer of electron sensitive resist. A grid of pixels is made on the substrate surface, each pixel having a unique address. The pattern data to be constructed is transferred to the computer. The computer controls the electron beam so as to create the pattern on the substrate pixel by pixel. The scanning beam deposits energy in the desired pattern on the resist film. Usually a Gaussian round electron beam or variable shaped electron beam is used for scanning. This technique can produce fine patterns because of the very small spot size of the electron beam.[2-4] Figure 4.2 depicts the basic diagram of an electron beam lithography setup.

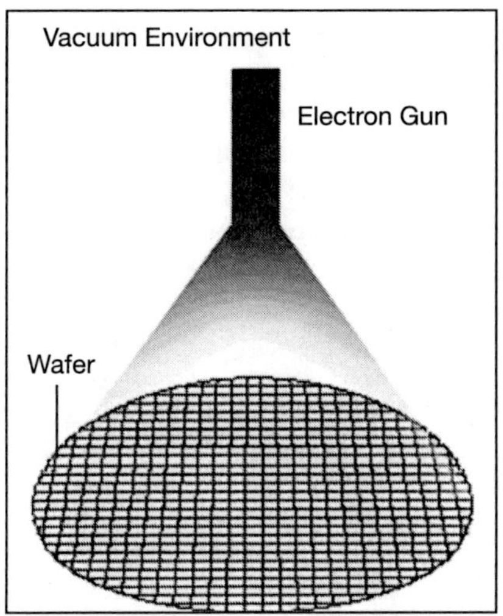

Fig. 4.2: Sketch of a typical electron beam lithography setup.

The diffraction effects are small in electron beam lithographic methods due to very small wavelength of electron beam. This technique has a large number of industrial applications like opto-electronic devices, optical devices, quantum structures and cryo-electric devices.

4.2.2 Nanoimprint Lithography (NIL)

Nanoimprint lithography technique is a non-conventional method for patterning of polymer nanostructures. NIL achieves pattern definition through direct mechanical deformation of the resist material. In this technique, a stamp or mould containing the desired pattern is pressed into a thin polymer film cast on a wafer at a controlled temperature and pressure. Schematic diagram of the pressing machine is shown in figure 4.3(a) and the various steps involved in the NIL process are shown in figure 4.3(b). The mould is made by producing a nanostructured surface in some wafer using electron beam lithography as described in section 4.2.1. Materials, such as silicon, silicon dioxide, metals and polymers can be used as stamps. Usually the stamp is coated with some anti-stiction coating to release the stamp easily after pressing.

Polymer materials are used as resists because above the glass transition temperature they are soft, make easy imprinting. Besides, polymers can be functionalized to become sensitive to various inputs like electric, magnetic, thermal, optical and biochemical input. The pressing machine provides necessary pressure, and maintains the temperature slightly above the glass transition temperature of the polymer.

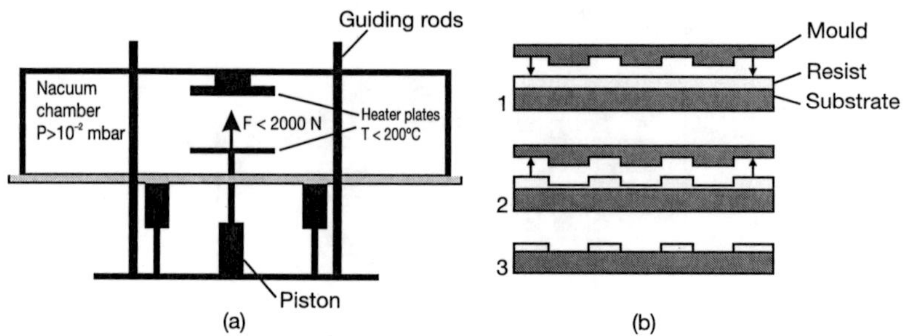

Fig. 4.3: (a) Schematic sketch of the pressing machine[4] and (b) sketch of the steps involved in NIL process: imprinting, after imprinting and after etching.

Source: www.azonano.com

Nanoimprint lithography technique is suitable for mass production. Once the mould is created, and the pressing machine is correctly arranged, it is simple to mass fabricate nanostructured wafers. The cycle time of a typical nanoimprint machine is of the order of minutes. This time scale is determined by the actual time it takes to press the stamp down and the various thermal time scales for heating and cooling of the sample. Nanoimprint can also be used to produce functional device structures in different polymers, which leads to an array of applications in biotechnology, photonics, data storage and electronics.

4.2.3 Dip-Pen Nanolithography (DPN)

Dip-pen nanolithography (DPN) technique transports ink on a sharp object to a paper substrate using capillary forces. This method has been used to deliver molecules in a positive printing mode.[5] DPN is an atomic force microscopy based technique developed by Chad Mirkin at Northwestern University. Dip-pen nanolithography delivers molecules from an AFM tip to a solid-state substrate via capillary transport. Moreover, DPN is a useful tool for generating and functionalizing nanodevices. Schematic representation of DPN technique is shown in figure 4.4.

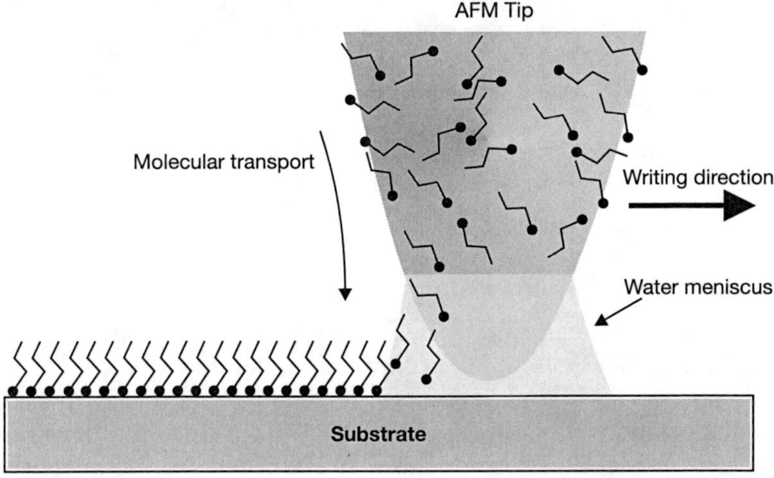

Fig. 4.4: Schematic diagram of a DPN process.

Source: www.npl.co.uk

During the contact mode operation of DPN, under ambient laboratory conditions, a water meniscus naturally forms between the ink-coated probe tip and the substrate. The ink moves on the substrate by capillary transport through the meniscus. One key issue of successful DPN patterning is choosing an ink and substrate with an appropriate chemical affinity. This causes the ink molecules to chemosorb onto the substrate. Proper binding self-regulates the diffusion of ink onto and across the substrate, thus controlling the resulting feature size and resolution. Figure 4.5 shows images of an array of dots and a 25 nm thick line, generated by DPN.

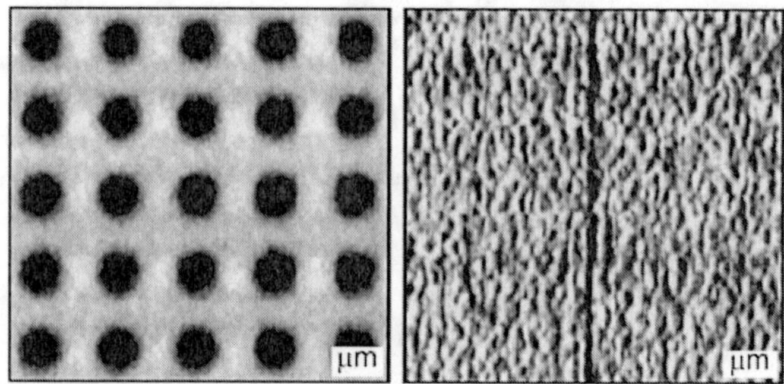

Fig. 4.5: Images of an array of dots, and a 25 nm thick line produced by DPN.

Source: www.eng.utah.edu

4.2.4 Microcontact Printing

Microcontact printing is a remarkable surface patterning technique. This stamping technique is developed by George Whitesides at Harward University. This technique can also transport molecules directly into substrates in a positive printing mode. Microcontact printing uses an elastomer stamp to deposit patterns of molecules directly to substrates. This technique helps to deposit the entire pattern on a substrate in a step, which is an advantage over DPN technique.[6]

The stamp is made by moulding polydimethylsiloxane (PDMS) on a master having the desired design. The master is produced by photolithographic techniques. Then the PDMS stamp can be inked with molecular compounds and applied to a surface such as thin film of gold as shown in figure 4.6. The molecular ink is left behind on the substrate and produce original master.

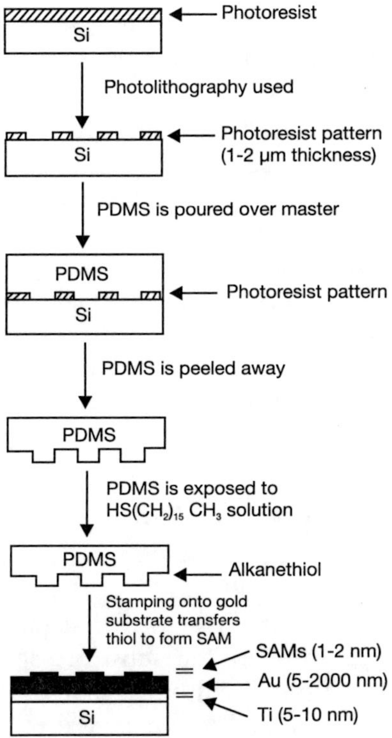

Fig.4.6: Schematic diagram of the procedure for microcontact printing.

Source: www.annualreviews.org

4.3 Epitaxy

Epitaxy is the method of depositing a thin film of single crystal material on a monocrystalline substrate. The film deposited is called epitaxial film or epitaxial layer. The term epitaxy is derived from the Greek words 'epi' means above and 'taxis' means 'in ordered manner'. So the word epitaxy means 'to arrange upon'. Epitaxial films can be grown from gaseous or liquid precursors, and the substrate acts as a seed crystal.[7] Also, orientation of the film deposited is identical to those of the substrate. The other thin-film deposition methods grow polycrystalline or amorphous films even on single-crystal substrates. When the film and substrate are same materials, the process is called homoepitaxy or simply isoepitaxy. If the film and substrate are different materials, the process is named heteroepitaxy.

The epitaxial growth techniques are widely used for the synthesis of nanostructured materials and semiconductors. This is an affordable method of high quality crystal growth for semiconductor materials like silicon, germanium, gallium nitride, gallium arsenide and indium phosphide. Epitaxy can also be used to deposit thin layers of pre-doped silicon on the polished sides of silicon wafers, before they are processed into semiconductor devices. There are different type of epitaxial growth techniques, such as vapour phase epitaxy, molecular beam epitaxy, liquid phase epitaxy, solid phase epitaxy and atomic layer epitaxy.[8]

4.3.1 Vapour Phase Epitaxy (VPE)

Vapour phase epitaxy is one of the most significant techniques for epitaxial growth. In VPE process, a number of gases are taken in an induction heated reactor where only the substrate is heated. The temperature of the substrate should be at least 50 per cent of the melting point of the material to be deposited. Typical sketch of a vapour phase epitaxial reactor is shown in figure 4.7.

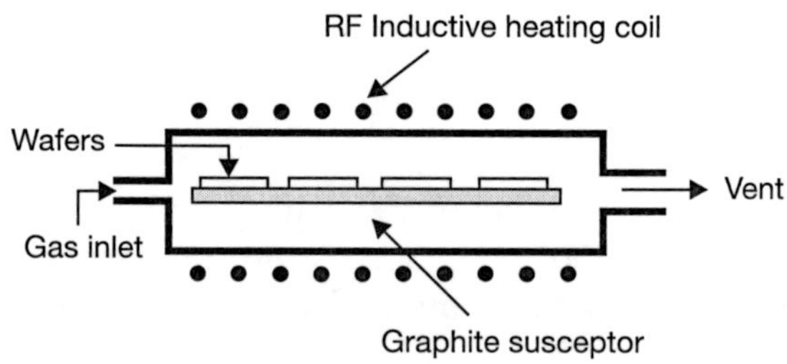

Fig. 4.7: Sketch of a typical VPE reactor.

Source: www.memsnet.org

Silicon from silicon tetrachloride is deposited in hydrogen atmosphere at about 1200°C. The reversible chemical reaction is

$$SiCl_{4(g)} + 2H_{2(g)} \rightarrow Si_{(s)} + 4HCl_{(g)} \quad ...(4.1)$$

The growth rate of the reaction depends on the proportion of the two source gases. If the growth rate is above 2 micrometres per minute, the reaction can produce polycrystalline silicon. But negative growth rates (etching) may occur if large hydrogen chloride byproduct is present. Besides, an additional etching reaction competes with the deposition reaction,

$$SiCl_{4(g)} + Si_{(s)} \rightarrow 2SiCl_{2(g)} \qquad ...(4.2)$$

Silane, dichlorosilane and trichlorosilane source gases are also used for silicon VPE. For example, the silane reaction occurs at 650°C, the reaction is

$$SiH_4 \rightarrow Si + 2H_2 \qquad ...(4.3)$$

This reaction occurs at lower temperatures than deposition from silicon tetrachloride. Also, it does not accidentally etch the wafer. However, it will form a polycrystalline film unless strongly controlled. VPE can be classified into hydride VPE and metal-organic VPE based on the source gases.

4.3.2 Liquid Phase Epitaxy (LPE)

LPE is a suitable technique to synthesise semiconductor crystal layers from the melt on solid substrates. This process occurs at a temperature much lower than the melting point of the deposit material. The material to be deposited is first dissolved in the melt of another material. At equilibrium conditions, the rate of deposition of the semiconductor crystal is slow and uniform. The equilibrium between dissolution and deposition depends on both the temperature and the concentration of the dissolved semiconductor in the melt. It is possible to control the growth of the layer from the liquid phase by cooling the melt. Besides, it is also possible to reduce impurity.

This epitaxial method is commonly used for the synthesis of compound semiconductors. For example, growth of ternary and quarternary III-V compounds on gallium arsenide (GaAs) or indium phosphide (InP) substrates. Glass or ceramic substrates can also be used for special applications. If the thermal expansion coefficient of substrate and the semiconductor layer is almost equal, the nucleation is smooth and the layer grown is tension free. The main advantage of this technique is that it can produce uniform and high quality thin layers.

4.3.3 Solid Phase Epitaxy (SPE)

In Solid Phase Epitaxial technique, a film of amorphous material is deposited on a crystalline substrate. Then the substrate is suitably heated to crystallize the film. The single crystal substrate serves as a template for crystal growth. It is also possible to use the annealing steps to recrystallize silicon layers amorphized during ion implantation, another type of Solid Phase Epitaxy. Impurity segregation and redistribution at the growing crystal-amorphous layer interface during this process are important research topics.

4.3.4 Molecular Beam Epitaxy (MBE)

Molecular beam epitaxy is an effective epitaxial growth technique to fabricate metal, insulator, semiconductor or superconductor epitaxial layers. MBE essentially consists of an ultra-high vacuum (UHV) chamber into which a substrate is loaded onto a heated sample holder. Precursors of desired elements (Ga, As, Al, P, In, etc.) are then loaded into heated crucibles or furnaces called Knudsen cells, outfitted with computer controlled shutters on their exits.[9, 10] The precursors are then heated such that when the shutters are opened, one obtains a beam of atoms directed towards the substrate. Under such low pressures, the atomic species have very long mean free paths allowing them to reach the substrate without collisions with other gas phase species in the chamber. By controlling the temperature as well as the sequence/timing of opening and closing the shutters, one can deposit uniform films of semiconductor materials. In this way one can obtain precise nanometre length scale quantum well structures.

Schematic representation of a typical MBE system is shown in figure 4.8. The source materials are placed in effusion cells to ensure an angular distribution of atoms or molecules in a beam. The substrate is sufficiently heated and frequently rotated to improve the growth homogeneity. A manipulator connected to the substrate holder controls position of the substrate relative to the effusion cells to get the required structures. The entire MBE chamber is kept at ultra high vacuum ($\sim 10^{-11}$ mbar) and its operation is oil free so that the substrate is clean before growing the film. Ultra high vacuum condition enables the molecular beam condition, where the geometrical size of the chamber is less than the mean free path of the particle.[11-14] Figure 4.9 shows the plot of mean free path of nitrogen molecules (300 K) against pressure.

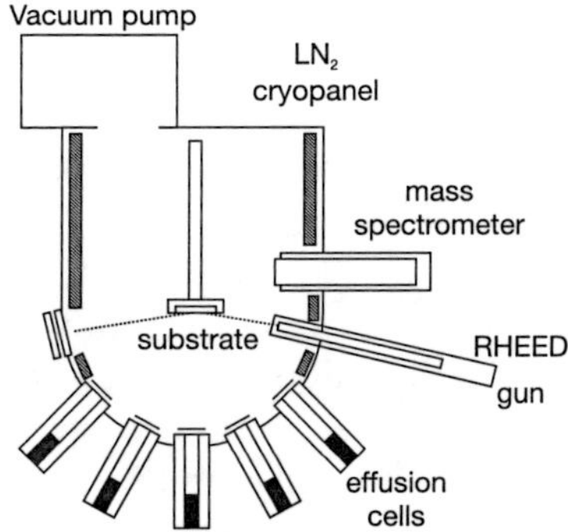

Fig. 4.8: Schematic diagram of a typical MBE system.

Source: www.wsi.tum.de

Generally, pyrolytic boron nitride (PBN) is chosen for the crucibles which gives low rate of gas evolution and chemical stability up to 1400°C. The shutters and heaters are made from molybdenum or tantalum. To create UHV, a bake out of the chamber at about 200°C for 24 hours is required.

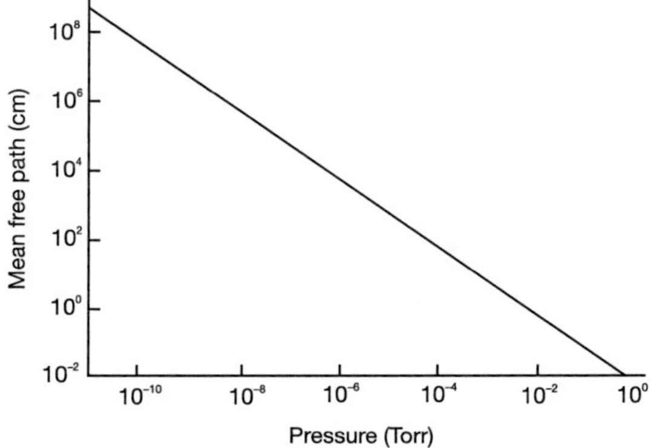

Fig. 4.9: The plot of mean free path of nitrogen molecules against pressure.

The atoms or molecules striking on the single crystal substrate result in the formation of the desired epitaxial film. Spurious fluxes of atoms and molecules in the chamber walls are reduced by a cryogenic screening around the substrate. The MBE systems also control the composition and doping of the growing structure at monolayer level. It is made possible by closing and opening shutters, and thereby changing the nature of the incoming beam. The operation time of a shutter (~ 0.1 s) is much smaller than the growing time for one monolayer (~ 1-5 s). The reflection high energy electron diffraction (RHEED) system and mass spectrometers attached to the MBE system are suitable for monitoring the beams and the molecular composition. The period of oscillation of the RHEED signal indicates the time required to grow a monolayer and the diffraction pattern obtained on the RHEED window gives the state of the surface as shown in figure 4.10.

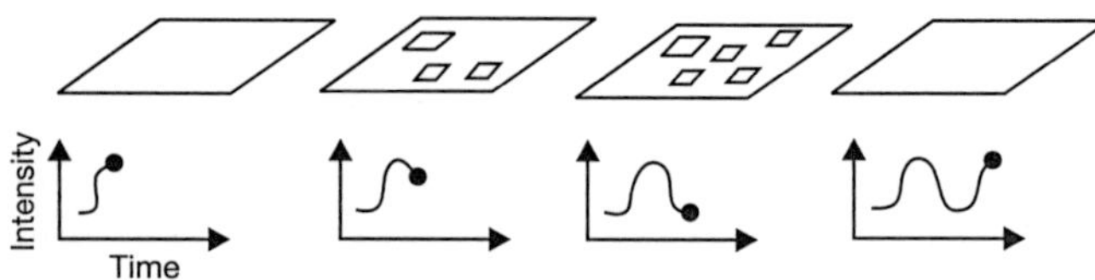

Fig. 4.10: RHEED oscillations.

Characteristics of MBE

Following are some of the important characteristics of MBE:

- Low growth rate of ~ 1 monolayer (lattice plane) per sec. A very smooth surface and interface is achievable through controlling the growth at the monoatomic layer level.
- Low growth temperature (~ 550°C for GaAs). A low growth temperature limits diffusion and maintains hyperabrupt interfaces, which are very important in fabricating two-dimensional nanostructures or multilayer structures such as quantum wells.
- Precise control of surface composition and morphology.
- Grow films with good crystal structure.

4.3.5 Atomic Layer Epitaxy (ALE)

Atomic layer epitaxial technique is suitable for depositing thin films and related surface structures.[15, 16] It has a self-limiting film growth mechanism. Simple and accurate film thickness control, uniformity over large areas, good reproducibility, excellent conformality, high film qualities and multilayer processing capability are some of the unique properties of ALE. The above properties make ALE technique suitable for nanotechnology. In the literature, ALE is also called atomic layer deposition (ALD), atomic layer growth (ALG), atomic layer CVD (ALCVD) or molecular layer epitaxy (MLE).

In ALE process, vapours of reactant are pulsed onto the substrate alternately. The reactor is purged with an inert gas between the intervals of the reactant pulses. The subsequent purging step removes all the unwanted molecules from the reactor chamber. When the next pulse is dosed in, it will form only the surface monolayer with which it reacts. The growth of ZnS film from $ZnCl_2$ and H_2S is shown in figure 4.11. Experimental conditions are properly adjusted to make all the process steps saturative. As shown in figure 4.11, each cycle has four steps and the product of one cycle is only a monolayer and very often a submonolayer. Thus, this technique is a time consuming growth process. However, proper reactor and precursor design can compensate this limitation. The self-limiting

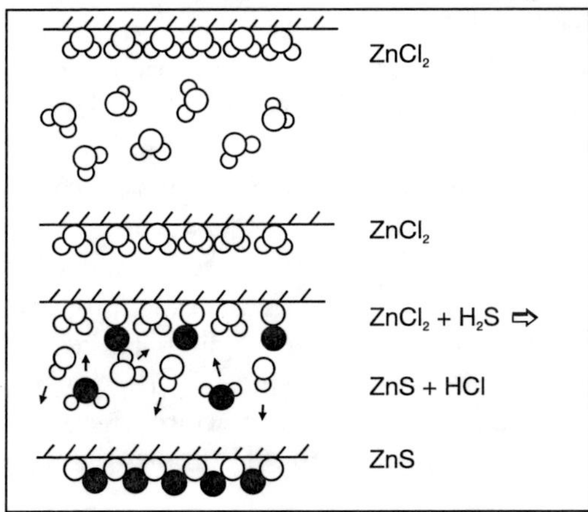

Fig. 4.11: Basic principle of the ALE process.

growth mechanism of ALE makes it ideal for thin film deposition. But the slowness of the film growth greatly limit the use of ALE. Flow-type reactors with closely spaced substrates and plug like flow conditions can finish each cycle within 0.1–0.2 s. It is possible to increase growth rate by making use of the large batch processing capability of ALE. For example, there are 82 glass substrates used in the largest ALE reactors for industrial manufacturing of thin film electroluminescent (TFEL) displays.[17]

The real process in ALE is a complicated one. In many cases, the incoming precursors react with the functional groups on the surface and there is no molecular chemisorption. The growth of metal oxides from metal chlorides and water is an example. For example, $TiCl_4$, binds to the surface by undergoing an exchange reaction with the surface hydroxyl groups:[18, 19]

$$n(\text{-OH})(s) + TiCl_4(g) \rightarrow (\text{-O-})_n TiCl_{4-n}(s) + n HCl(g) \quad ...(4.4)$$

where n = 1–3, depends on the temperature and spatial distribution of the hydroxyl groups. The following water pulse changes the surface back to a hydroxylated one:

$$(\text{-O-})_n TiCl_{4-n}(s) + (4-n)H_2O(g) \rightarrow (\text{-O-})_n Ti(OH)_{4-n}(s) + (4-n)HCl(g) \quad ...(4.5)$$

If the precursor doses and pulse times are sufficient for reaching the saturated state and no extensive precursor decomposition takes place, the ALE exhibits excellent conformality. Besides the films fabricated by ALE is free from point defects due to high conformality. This is significant in depositing high quality insulators for TFEL devices and dense corrosion protection coatings. The conformality requirements in processing porous substrates, like porous silica and alumina powders and porous silicon layers have been satisfied by the ALE deposition. Since the time required to transport precursors into and out of the nanometre size pores is long, usually only a few ALE deposition cycles can be applied. Therefore, these ALE processes help to modify the surfaces of the porous substrates rather than to deposit a thin film onto them. Moreover, a uniform distribution through the nanoporous substrate can be confirmed.

Film thickness control in ALE can be made simple because during each deposition cycle exactly the same quantity of material is deposited. Epitaxial superlattice is an example for monolayer accuracy in the film thickness control. In the deposit of superlattices of the type $(InAs)_1(GaAs)_5$ one InAs monolayer is formed between five GaAs monolayers.

In the preparation of nanolaminate dielectric films, accurate thickness control of ALE is of great importance. Such nanolaminate films are suitable to develop insulators with high permittivity and low leakage current. Usually, Ta_2O_5 possesses relatively high permittivity of about 25, but it is very leaky. In the nanolaminates, Ta_2O_5 is combined with other insulators like ZrO_2 or HfO_2, into a stack of thin (2.5–15 nm) layers to produce a drastic improvement in leakage current properties with a small decrease of the permittivity. In the nanolaminates, localized weak points in one sublayer are counterbalanced by the adjacent layers. This property restricts the leaky channels to extend the whole nanolaminate structures. The accurate film thickness control and large area uniformity of the ALE makes it best suited in achieving the optimized dielectric properties.

4.4 Sputtering

In sputtering, energetic ions are used to knock atoms or molecules out from a target that acts as one electrode and subsequently deposit them on a substrate acting as another electrode.[20] Figure 4.12 schematically illustrates the principles of dc or RF sputtering systems. Target and substrate serve as electrodes and face each other in a typical sputtering chamber. An inert gas, typically argon with a pressure usually ranging from a few to 100 mTorr, is introduced into the system as the medium to initiate and maintain a discharge. When an electric field of several kilovolts per centimetre is introduced, a glow discharge is initiated and maintained between the electrodes. The free electrons are accelerated by the electric field, and gain sufficient energy to ionize argon atoms. The gas density or pressure must not be too low, or else the electrons will simply strike the anode without having gas phase collision with argon atoms. However, if the gas density or pressure is too high, the electrons will not have gained sufficient energy when they strike gas atoms to cause ionization. The positive ions, Ar^+, produced in the discharge strike the cathode (the source target) resulting in the ejection of neutral target atoms through momentum transfer. These ejected atoms move towards the opposite electrode and deposit there (the substrate with growing film). In addition to the growth species, i.e. neutral atoms, other negatively charged species under the electric field will also bombard and interact with the surface of the substrate or grown film.

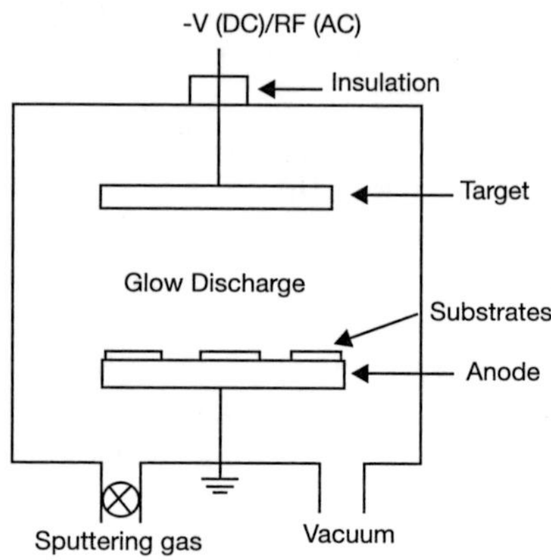

Fig. 4.12: Schematic diagram of the sputtering system.

Source: www.wizam.eu

For the deposition of insulating films, an alternate electric field is applied to generate plasma between two electrodes. Typical RF frequencies employed range from 5 to 30 MHz. The key element in RF sputtering is that the target self biases to a negative potential and behaves like a dc target. To prevent simultaneous sputtering on the grown film or substrate, the sputter target must be an insulator and be capacitively coupled to the RF generator. This capacitor will have low RF impedance and will allow the formation of a dc bias on the electrodes.

4.5 Chemical Vapour Deposition (CVD)

CVD is a chemical process that is used to deposit high purity solid materials in monocrystalline, polycrystalline, amorphous and epitaxial forms.[21-23] Materials such as silicon, carbon fibre, carbon nanofibres, filaments, carbon nanotubes, SiO_2, silicon-germanium, tungsten, silicon carbide, silicon nitride, silicon oxynitride, titanium nitride and various high k dielectrics are examples of solid materials for chemical vapour deposition. Microfabrication and nano-fabrication processes are widely used CVD to deposit materials. Besides, this process is also used to produce thin films for semiconductor fabrication, and synthetic diamonds. The substrate (wafer) is exposed to one or more volatile precursors in a typical CVD process. The precursors react and/or decompose on the substrate surface to produce the deposit. The volatile by-products produced during the process are removed by gas flow through the reaction chamber.

The CVD process is complicated and involves a number of gas-phase and surface reactions.[24-30] They are often summarized by overall reaction schemes, as illustrated in scheme 4.1.

$$[Me_3Ga]_{(g)} + [AsH_3]_{(g)} \rightarrow GaAs_{(S)} + 3CH_{4(g)} \uparrow$$
$$[SiH_4]_{(g)} \rightarrow Si_{(S)} + 2H_{2(g)} \uparrow$$
$$[SiH_4]_{(g)} + O_{2(g)} \rightarrow SiO_{2(S)} + 2H_{2(g)} \uparrow$$
$$2[Ta(OEt)_5]_{(V)} + 5H_2O_{(V)} \rightarrow Ta_2O_{5(S)} + 10EtOH_{(V)} \uparrow$$
$$[NbCl_5]_{(v)} + 5/2H_{2(g)} + {}^1/2xN_{2(g)} \rightarrow NbN_{x(S)} + 5HCl_{(V)} \uparrow$$

Scheme 4.1: Overall reaction schemes for variety of CVD processes.

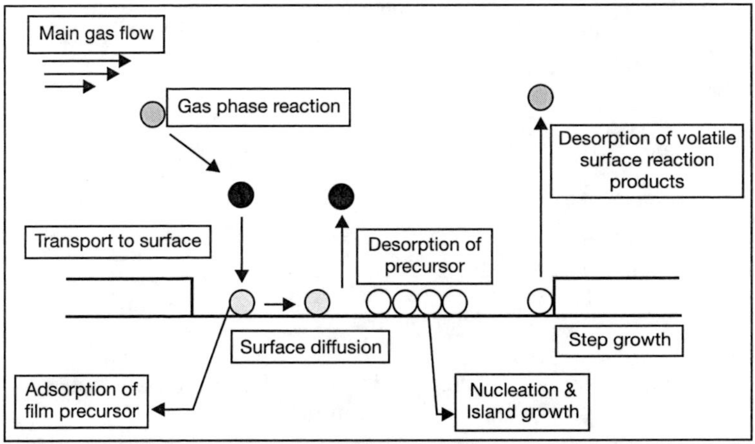

Fig. 4.13: Precursor transport and reaction processes in CVD.

A detailed picture of the basic physico-chemical steps in an overall CVD reaction is illustrated in figure 4.13, which indicates several key steps:

- Evaporation, and transport of precursors to the reactor;
- Gas phase reactions of precursors;
- Mass transport of reactants to the substrate surface;
- Adsorption of the reactants on the substrate surface;
- Surface diffusion, nucleation and surface chemical reactions leading to film formation;
- Desorption of precursors and volatile by-products.

In traditional thermal CVD, the film growth rate is determined by several parameters, such as pressure of the reactor, the temperature of the substrate and nature of the gas-phase.

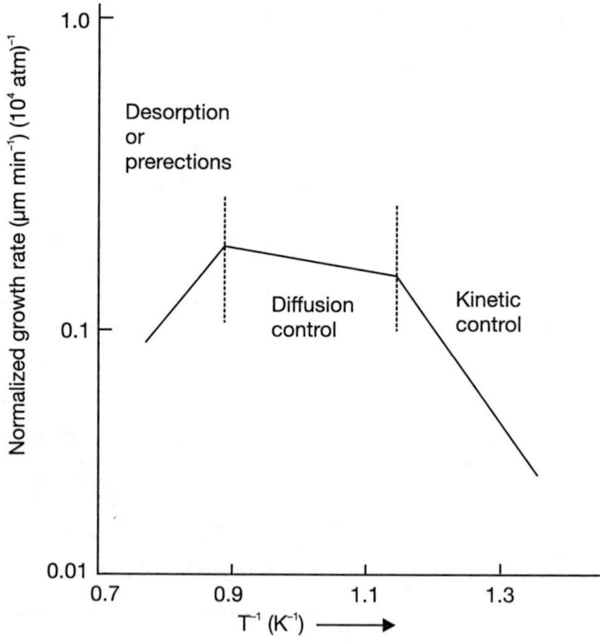

Fig. 4.14: The normalized growth rate of GaAs as a function of temperature.

The graph of the normalized CVD growth rate of GaAs as a function of growth temperature is shown in figure 4.14. In this plot of log of growth rate versus reciprocal thermodynamic temperature, three regions are apparent. At low temperatures, the growth rate is controlled by the kinetics of chemical reactions in the gas-phase or on the substrate surface. This region is called as kinetic growth control region. The growth rate of the film increases exponentially with substrate temperature. According to the Arrhenius equation,

$$\text{Growth rate} \propto e^{(E_A/RT)} \qquad ...(4.6)$$

where E_A is the apparent activation energy, R the gas constant and T the temperature.

When the temperature increases, the growth rate is almost independent of temperature. But, it is controlled by the mass transport of reagents to the growth surface. This region is called mass transport or diffusion-controlled growth region. The growth rate shows a decrease at higher temperatures because of increased rate of desorption of precursors from the growth surface. Gas-phase reactions become increasingly important with increasing temperature and higher partial pressures of the reactants.

The pressure of the CVD reactor has significant role in determining the relative importance of each regime. The gas phase reactions are important from atmospheric pressure (760 Torr) to intermediate pressure (e.g. 10 Torr). Both kinetics and mass transport can play an important role in deposition. As the pressure falls below 1 Torr, layer growth is often controlled by surface reactions. There is no mass transport at very low pressures. At very low pressures, the layer growth is often controlled by the temperatures of gas and substrate, and by desorption of precursor fragments and matrix elements from the growth surface.

4.5.1 Types of Chemical Vapour Deposition Process

There are various forms of CVD processes and are different in the way by which chemical reactions are initiated and process conditions.

Classification based on operating pressure

(a) Atmospheric pressure CVD (APCVD): In APCVD, chemical vapour deposition processes take place at atmospheric pressure.

(b) Low-pressure CVD (LPCVD): In LPCVD, chemical vapour deposition processes take place at sub-atmospheric pressures. The reduced pressures can reduce unwanted gas-phase reactions. Hence, the film uniformity across the wafer will be improved.

(c) Ultrahigh vacuum CVD (UHVCVD): As the name implies, chemical vapour deposition processes occur at ultrahigh vacuum in UHVCVD, below 10^{-6} Pa ($\sim 10^{-8}$ Torr).

Classification based on physical characteristics of vapour

(a) Aerosol assisted CVD (AACVD): In AACVD process, the precursors are transported to the substrate using a liquid or gas aerosol, which will be generated ultrasonically.

(b) Direct liquid injection CVD (DLICVD): In DLICVD process, the precursors are in liquid form and are injected in a vapourization chamber towards injectors. Then the precursor vapours are transported to the substrate. This technique is suitable for both liquid and solid (dissolved in suitable solvent) precursors. The direct liquid injection CVD technique offers high growth rates.

Classification by the source of energy supplied for the chemical reaction enhancement inside the CVD chamber

(a) Plasma enhanced CVD (PECVD): In PECVD process, plasma is generated to enhance chemical reaction rates of the precursors. This technique permits deposition even at lower temperatures, which helps the manufacture of nano-semiconductor materials. PECVD process can be used to deposit thin films on a substrate. Chemical reactions occur after creation of plasma of the reacting gases. The plasma is usually generated by RF (AC) frequency or DC discharge between two electrodes. The space between the electrodes is filled with the reacting gases.

(b) Remote plasma-enhanced CVD (RPECVD): In RPECVD technique, the wafer substrate is not directly in the plasma discharge region. As the wafer removes from the plasma region, the processing temperatures come down to room temperature.

(c) Atomic layer CVD (ALCVD): This technique deposits successive layers of different substances to produce layered, crystalline films.

(d) Hot wire CVD (HWCVD): In HWCVD process, a hot filament supplies sufficient energy to chemically decompose the source gases.This technique is also called as catalytic CVD (Cat-CVD) or hot filament CVD (HFCVD). HWCVD process is often used for the synthesis of various nanomaterials.

(e) Rapid thermal CVD (RTCVD): Rapid thermal CVD process use heating lamps or other techniques to rapidly heat the wafer substrate. But, it cannot heat the gas or chamber walls. Hence, it reduces the unwanted gas phase reactions and particle formation.

(f) Metal-Organic CVD (MOCVD): In this technique, CVD process is based on metalorganic precursors.

4.5.2 Metal-Organic Chemical Vapour Deposition (MOCVD)

MOCVD can be used to grow thin films of materials, like molecular beam epitaxy. This technique can also be used for the synthesis of semiconductor nanowires in the presence of gold nanoparticle catalysts.[31, 32] The principle of operation of MOCVD is similar to that of MBE. The schematic diagram of a typical MOCVD system is shown in figure 4.15. Usually a MOCVD set up consists of a quartz tube furnace with a heated substrate called reactor, gas handling system, vacuum and exhaust system, and control system.

Organo metallic or metal-organic compounds such as trimethylaluminium, trimethylgallium, trimethylindium as well as gases such as phosphine or arsine can be introduced into the heated reactor and allowed to decompose giving the required elemental species. The vapour pressure of the metal organic source is an important consideration in MOCVD, since it determines the concentration of source material in the reactor and the deposition rate. Too low a vapour pressure makes it difficult to transport the source into the deposition zone and to achieve reasonable growth rates. Very high vapour pressure may raise safety concerns if the compound is toxic. Further more, it is easier to control the delivery from a liquid than from a solid.

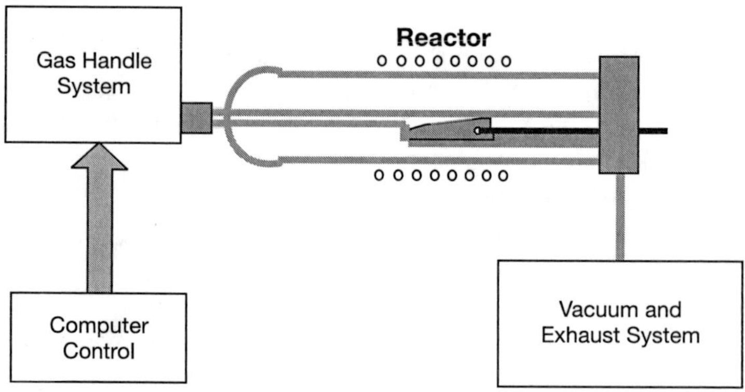

Fig. 4.15: Typical MOCVD growth system.

Gas Handling System

The function of gas handling system is mixing and metering of the gas that will enter the reactor. Timing and composition of the gas entering the reactor will determine the epilayer structure. Leak-tight of the gas panel is essential, because the oxygen contamination will degrade the properties of growing films.

Vacuum and Exhaust System

It consists of pump and pressure controller, and waste gas treatment system. For low pressure growth, we use mechanical pump and pressure controller to control the growth pressure. The pump should be designed to handle large gas load. The treatment of exhaust gas is a matter of safety concern. The MOCVD system for GaAs and InP use toxic materials like AsH_3

and PH_3. The exhaust gases still contain some unused AsH_3 and PH_3. Normally, the toxic gas need to be removed by using chemical scrubber. Comparison of different epitaxial techniques, including their salient features and limitations, are presented in Table 4.1.

Table 4.1: Comparison of different epitaxial methods

Growth method	Year	Features	Limit
LPE	1963	Growth from supersaturated Solution onto substrate	Limited substrate areas and poor control over the growth of very thin layers
VPE	1958	Use metal halide as transport agents to grow	No Al contained compound; thick layer
MBE	1967	Deposit epilayer at ultra-high vacuum	Hard to grow materials with high vapour pressure
MOCVD	1968	Use metal organic compounds as the source	Some of the sources like AsH_3 are very toxic

4.6 Sol-Gel Technology

Sol-gel technology is a versatile solution process for synthesising ceramic and glass materials. The sol-gel technique involves the transition of a liquid "sol" (colloidal) into a solid "gel" phase. This technology help us to make ultra-fine powders, thin films, ceramic fibres, extremely porous aerogels and so on.[33, 34] Schematic diagram of a sol-gel process is shown in figure 4.16.

Nowadays, the sol-gel process is quite often used for the fabrication of a variety of nanomaterials. The sol-gel process initiates from a chemical solution, which acts as the precursor for an integrated network (gel) that reacts to form nanosized colloidal particles (sol). Usually, metal alkoxides and metal chlorides are used as precursors. They undergo different hydrolysis and polycondensation reactions. Metal oxo or metal hydroxo polymers in a solution can be generated by connecting the metal centres with oxo (M-O-M) or hydroxo (M-OH-M) bridges. Thus, the sol evolves towards the formation of a gel like diphasic system containing both a liquid phase and solid phase whose morphologies range from discrete particles to continuous polymer networks.

The particle density is low in a colloid. Hence, large quantity of fluid may be removed from it to recognize the gel like properties. A simple method is to allow time for sedimentation to occur and pour off rest of the liquid. Centrifugation method can also be used for the process of phase separation. A drying process can remove off the remaining liquid (solvent), which results in shrinkage and densification. One of the unique advantages of using this technique is that it ensures densification even at a very low temperature.

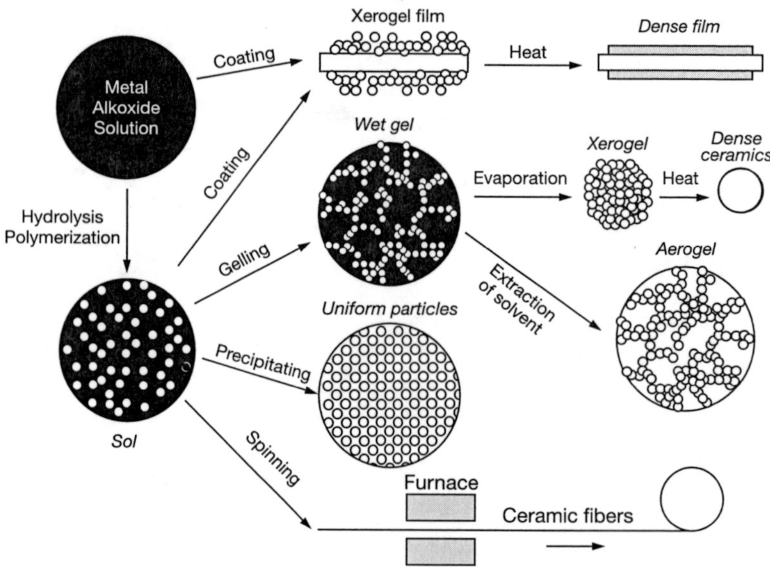

Fig. 4.16: Schematic diagram of a sol-gel process.

Source: www.gryfinska.pl

The precursor sol can be deposited on a substrate to form a film (e.g. dip-coating or spin-coating). It is also possible to use the sol to synthesize powders (e.g. microspheres and nanospheres). When the liquid in a wet "gel" is removed under a supercritical condition, a highly porous and extremely low density "aerogel" will be obtained. If the viscosity of a "sol" is adjusted into a suitable range, ceramic fibres can be drawn.

Ultra-fine and uniform ceramic powders can be synthesized by precipitation, spray pyrolysis or emulsion techniques.

The low temperature sol-gel technique has fine control over the chemical composition of products. The dopants added to the sol (e.g. organic dyes and rare earth metals) can be uniformly dispersed in the final product. Sol-gel derived materials have diverse applications in various fields like optics, electronics, energy, space, biosensors, medicine (e.g. controlled drug delivery), and so on.

4.6.1 Polymerization of Metal Alkoxides

Metal alkoxides (R-O-M) are organic compounds with one or more metal atoms in the molecule. They are like alcohols (R-OH) with a metal atom M, replacing the hydrogen H in the hydroxyl group. Metal alkoxides are widely used in sol-gel synthesis as ideal chemical precursors.[35-37]

Silicon tetraethoxide, or tetraethyl orthosilicate (TEOS) is an example for metal alkoxide. Its chemical formula is $Si(OC_2H_5)_4$ or $Si(OR)_4$, where the alkyl group $R = C_2H_5$. Metal alkoxides react readily with water, and a hydroxyl ion becomes attached to the metal atom during the reaction. This reaction is called hydrolysis.

$$Si(OR)_4 + H_2O \rightarrow HO\text{-}Si(OR)_3 + R\text{-}OH$$

During the completion of hydrolysis process, all of the OR groups are replaced by OH groups, depending on the amount of water and catalyst present.

$$Si(OR)_4 + 4H_2O \rightarrow Si(OH)_4 + 4R\text{-}OH$$

Any intermediate species $[(OR)_2\text{–}Si\text{-}(OH)_2]$ or $[(OR)_3\text{–}Si\text{-}(OH)]$ will be considered the result of partial hydrolysis. Besides, as two partially hydrolysed molecules can link together in a condensation reaction to form a siloxane [Si–O–Si] bond:

$$(OR)_3\text{–}Si\text{-}OH + HO\text{–}Si\text{-}(OR)_3 \rightarrow [(OR)_3Si\text{–}O\text{–}Si(OR)_3] + H\text{-}O\text{-}H$$

or

$$(OR)_3\text{–}Si\text{-}OR + HO\text{–}Si\text{-}(OR)_3 \rightarrow [(OR)_3Si\text{–}O\text{–}Si(OR)_3] + R\text{-}OH$$

Thus, polymerization of metal alkoxides is associated with the formation of a 1, 2, or 3-dimensional network of siloxane

[Si–O–Si] bonds along with the production of H-O-H and R-O-H species. This will give the required sol. Then, the sol is aged for a certain time to get long chains of siloxane bonds, removed by hydroxyl groups, resulting in thickening of the sol to form the gel. Finally, this gel is heated up in atmospheric air to get SiO_2. The condensation liberates a small molecule, such as water or alcohol. This type of reaction can continue to build larger and larger silicon-containing molecules by the process of polymerization.

4.7 Colloidal Growth

Colloidal growth is a chemist's approach to make nanostructures.[38] There are different approaches for the colloidal synthesis. A colloidal synthesis system consists of precursors, surfactants and solvents. When the reaction medium is heated to a sufficiently high temperature, the precursors transform into active atomic or molecular species (monomers) and subsequently grow nanocrystals. Thus, the formation of nanocrystals involves nucleation of an initial seed and subsequent growth. In the nucleation step, precursors decompose or react at a suitable temperature to form a supersaturation of monomers followed by a burst of nucleation of nanocrystals. These nuclei then grow further by consuming additional monomers. The presence of surfactant molecules may help the subsequent growth of the nanocrystals. By varying the temperature and duration of heat treatment one can vary the average size of the nanocrystals.

The selection of suitable temperature for growth is important because the degree of hotness controls the rearrangement of atoms and anneal during growth. There is a reduction in melting temperature for crystals of smaller size. The reduction in melting temperature increases the range of inorganic colloidal nanocrystals, which can be grown at temperatures (200-400°C) where common organic molecules are stable.

To stabilize a colloid in the small cluster size regime, it is necessary to use some stabilizers that can attach to the cluster surface and thereby prevent the uncontrolled growth into larger particles. A more common approach is the use of a polymeric surfactant/stabilizer which is added to a reaction designed to precipitate the crystal. The polymer attaches (electrostatically) to the surface of the growing clusters, and prevents their further growth. The most commonly used polymer surfactant is sodium polyphosphate (hexametaphosphate). Clusters of CdS, CdTe and ZnTe can be prepared with sodium polyphosphate surfactant.

4.8 Nanotube Synthesis

Arc discharge, laser ablation and chemical vapour deposition are the three important fabrication techniques employed for the synthesis of carbon nanotubes.[39] More economic synthesis routes to produce these structures are under investigation. In arc discharge method, an arc discharge between two carbon electrodes with or without catalyst can create a carbon vapour. Nanotubes self-assemble from the resulting carbon vapour. In the laser ablation technique, a high power laser beam is incident on a volume of carbon containing feedstock gas like methane or carbon monoxide. This technique produces a small quantity of clean nanotubes, whereas arc discharge methods generally produce large amount of impure material. The chemical vapour deposition technique produces MWNTs or poor quality SWNTs. The SWNTs synthesized by CVD process have a large diameter range, which can be easily controlled. Moreover, the CVD method is very easy to scale up, that favours commercial production.

4.8.1 Arc Discharge

The carbon arc discharge is a simple method to produce carbon nanotubes. However, it produces a mixture of components and requires proper method to separate nanotubes from the soot and the catalytic metals present in the crude product. This method generates carbon nanotubes through arc-vapourisation of two carbon electrodes, separated by approximately 1 mm.[40-42] The system is placed in an enclosure which is filled with inert gas (helium/argon) at low pressure (between 50 and 700 mbar). A very high temperature is acquired which allows the sublimation of the carbon. The same technique can also be used in synthesizing fullerenes. It is also possible to create nanotubes with the arc discharge method in liquid nitrogen. A d.c. of 50 to 100 A derived from about 20 V creates a high temperature discharge between the two electrodes. The discharge vapourises one of the carbon rods and results in a small rod shaped deposit on the other rod. Producing nanotubes in high yield depends on the uniformity of the plasma arc and the temperature of the deposit formed on the carbon rod. The carbon nanotubes can be purified by gasification with oxygen or carbon dioxide. It is possible to selectively grow SWNTs or MWNTs with arc discharge apparatus, which is shown in figure 4.17.

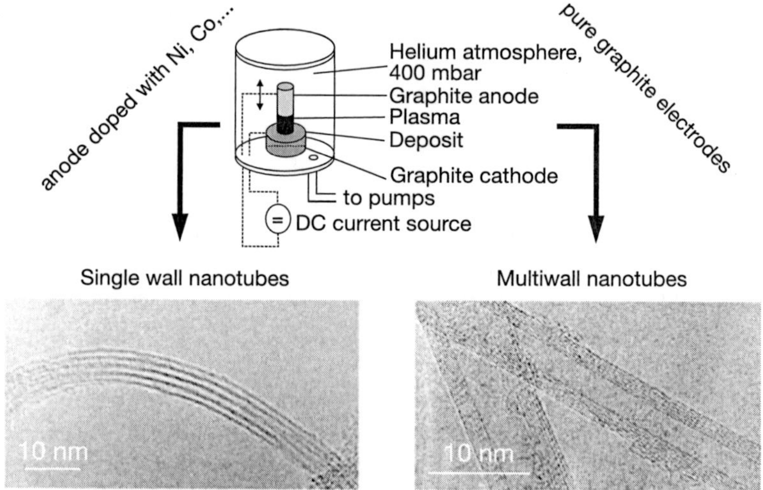

Fig. 4.17: Experimental set up of an arc discharge apparatus.

Source: www.students.chem.tue.nl

Synthesis of SWNT

In order to grow SWNTs, the anode of the arc discharge apparatus has to be doped with metal catalyst like Fe, Co, Ni, Y or Mo. A large number of elements and mixtures of elements have been tested by various researchers.[43] It has been observed that the results vary a lot, even with the same elements. The reason is that experimental conditions differ. Experimental parameters like metal concentration, inert gas pressure, kind of gas, current and system geometry determine the quantity and quality of the nanotubes. Its diameter is usually in the range of 1.2 to 1.4 nm.

The major problems associated with SWNT synthesis through arc discharge apparatus are (i) the product contains a lot of metal catalyst, (ii) SWNTs have defects, and (iii) purification is hard to perform. But it is possible to control the diameter of the nanotube by changing thermal transfer and diffusion. It has been reported that argon with a lower thermal conductivity and diffusion coefficient gives SWNTs with a smaller diameter of about 1.2 nm.

The plasma can be controlled by adjusting the anode to cathode distance. The anode to cathode distance is adjusted such that strong visible vortices will be around the cathode. This can improve anode vapourisation and thereby increase nanotubes formation. Catalysts also play a significant role in the control of the diameter of SWNTs. It has been reported that the diameter can be lowered to a range of 0.6-1.2 nm using suitable catalysts.

A modified arc discharge method using a bowl-like cathode is shown in figure 4.18. This new method can reduce defects, grow cleaner nanotubes and improve oxidation resistance. Strong oxidation resistance nanotubes are required for applications like field emission displays.

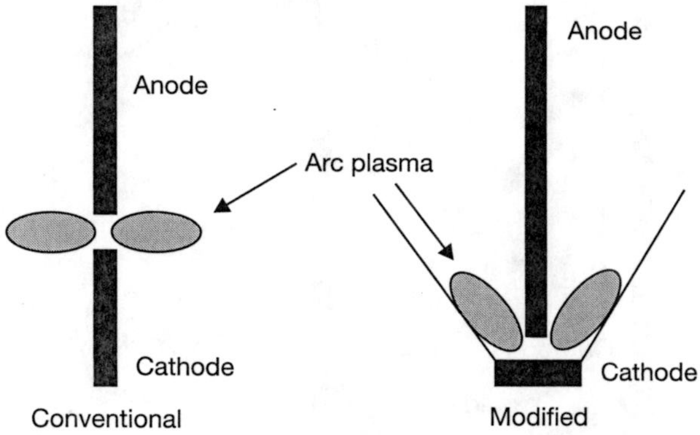

Fig. 4.18: Arc discharge electrode set-up: (a) the conventional and (b) the modified.

Source: www.students.chem.tue.nl

Synthesis of MWNT

When both the electrodes of arc discharge apparatus are made from graphite, the result will be MWNTs. But there are some side products such as fullerenes, amorphous carbon and some graphite sheets. There is no need of any heavy acidic purification steps because no catalyst is added to the reaction.

Synthesis in Liquid Nitrogen

Arc-discharge method in liquid nitrogen is an economical synthesis route for mass production of highly crystalline MWNTs. This method need not require low pressures and expensive inert gases. Moreover, around 70 per cent of the reaction products will be MWNTs.

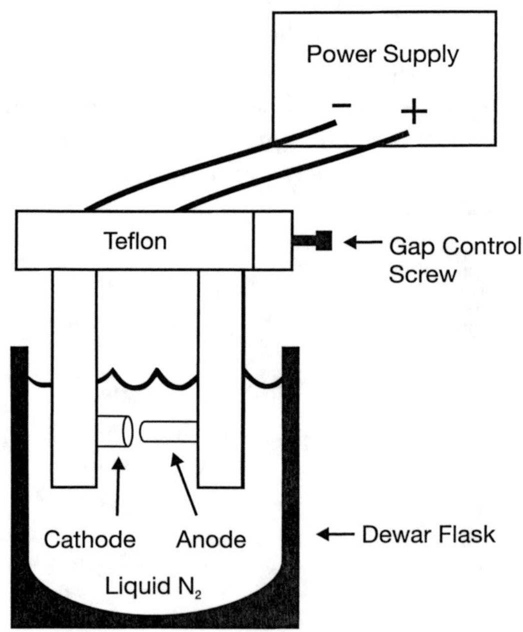

Fig. 4.19: Schematic diagram of the arc discharge apparatus in liquid nitrogen.

Source: www.students.chem.tue.nl

Magnetic Field Synthesis

Magnetic field synthesis routes of MWNTs generate less defective and high purity nanotubes. They can be used as nanosized electric wires for device fabrication. In this technique, a magnetic field around the arc plasma controls the arc discharge synthesis, which is shown in figure 4.20(a) and (b). Graphite rods with purity better than 99.99 per cent are used as electrodes. High purity MWNTs with purity greater than 95 per cent will be fabricated even without any purification steps. SEM images of MWNTs synthesized with and without the magnetic field are shown in figure 4.21(a) and (b).

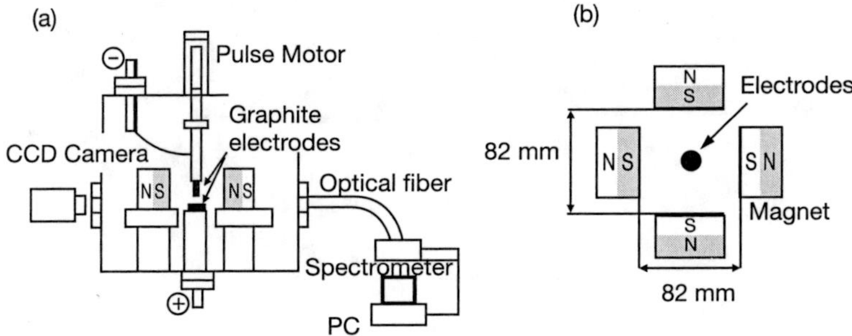

Fig. 4.20: Experimental set up for the synthesis of MWNTs in a magnetic field.

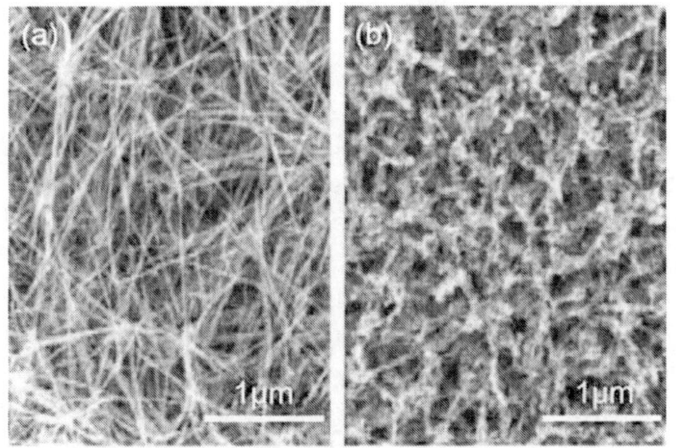

Fig. 4.21: SEM images of MWNTs: Synthesized (a) with magnetic field, and (b) without the magnetic field.

Source: www.students.chem.tue.nl

Plasma Rotating Arc Discharge

Plasma rotating arc discharge technique is another economical route to mass production of MWNTs. The experimental set-up of plasma rotating arc discharge technique is shown in figure 4.22. The centrifugal force due to the rotation creates turbulence and accelerates the carbon vapour perpendicular to the anode. Besides, the rotation uniformly distributes the micro discharges and creates stable plasma. As a result, volume and temperature of the plasma increases. By controlling the speed of rotation, one can vary the yield, even without any catalyst.

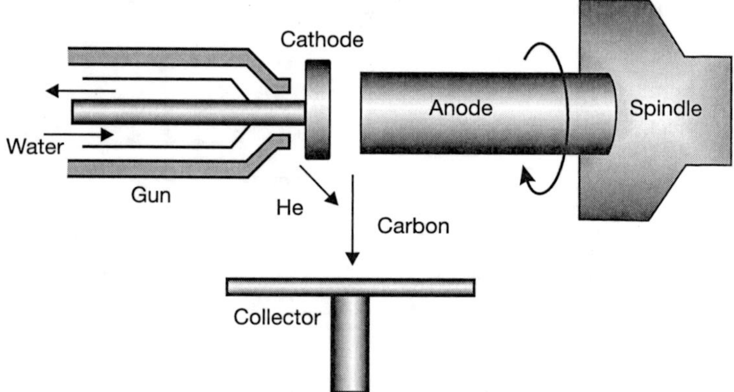

Fig. 4.22: Schematic diagram of plasma rotating electrode system.

Source: www.kazuli.com

4.8.2 Laser Ablation

Smalley's group (1955) at Rice University investigated laser ablation technique for the synthesis of carbon nanotubes.[44-50] This process ensures carbon nanotubes high purity single walls. Experimental set-up of a laser ablation system is shown in figure 4.23. A continuous or pulsed laser beam is made incident on a graphite target in a furnace at 1200°C. The laser beam vapourises the graphite target. Helium or argon gas is taken inside the furnace in order to keep the pressure at 500 Torr. The hot vapour plume expands and cools rapidly. When it cools, small carbon atoms and molecules condense to form larger clusters like fullerenes. The catalysts also begin to condense slowly and attach to carbon clusters, and prevent their closing into cage structures. Catalysts may even open cage structures when they attach to them. From these initial clusters, tubular molecules grow into single wall carbon nanotubes. The SWNTs formed are bundled together by van der Waals forces.

Laser ablation results in a higher yield for SWNT synthesis. The nanotubes produced by this method have better properties and a narrower size distribution compared to SWNTs produced by arc discharge. Nanotubes fabricated by laser vapourisation are purer than those created in the arc discharge method. For this process, a purification step by gasification is also required to eliminate carbonaceous material.

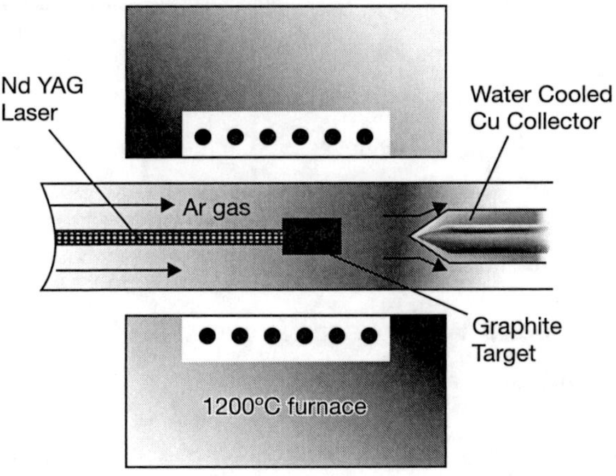

Fig. 4.23: Schematic drawings of a laser ablation apparatus.

Source: www.students.chem.tue.nl

4.8.3 Free Electron Laser (FEL) Method

The pulse width of a free electron laser system is ~ 400 fs and its repetition rate is 10-75 Hz. The pulse is focused into the graphite rod using a lens. The intensity of the laser bundle behind the lens will be ~5×10^{11} W/cm.[46] A jet of hot (1000°C) argon gas is passed through a nozzle tip that is situated close to the rotating graphite target, which contains the catalyst. The argon gas deflects the ablation plume nearly 90° away from the incident FEL beam direction. The SWNT soot thus produced is collected in a cold finger. The experimental set-up of a free electron laser process is shown in figure 4.24. The maximum possible yield with FEL process is 45 g/h, using NiCo or NiY catalyst. The SWNTs produced in bundles of 8-200 nm with a length of 5-20 microns have a diameter range 1-1.4 nm.

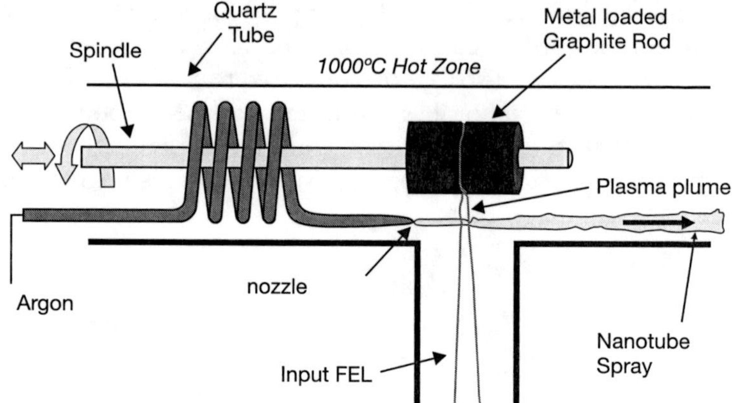

Fig. 4.24: Schematic diagram of a free electron laser apparatus.

Source: www.students.chem.tue.nl

4.8.4 Continuous Wave Laser Powder Method

Continuous wave laser powder method is a highly productive SWNT synthesis route.[48] In this system, a 2 kW continuous wave CO_2 laser in an argon or nitrogen stream is made incident on a mixture of graphite and metallic catalyst powders. Because of

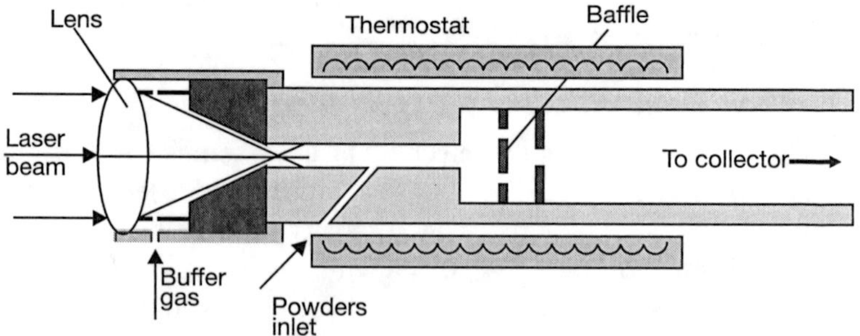

Fig. 4.25: The set-up for SWNT production by continuous wave laser powder method.

Source: www.students.chem.tue.nl

the introduction of micron size particle powders, thermal conductivity losses are significantly decreased compared with laser heating of the bulk solid targets in known laser techniques. As a result, more effective utilisation of the absorbed laser power for material evaporation is achieved. The set-up of the laser apparatus is shown in figure 4.25.

The established yield of this technique is 5 g/h. A Ni/Co mixture (Ni/Co is 1:1) is used as catalyst at a temperature of about 1100°C. In the soot, a SWNT abundance of 20-40 per cent can be formed with a mean diameter of 1.2-1.3 nm.

4.8.5 CVD Nanotube Synthesis

In CVD nanotube synthesis, a gaseous carbon source (methane, carbon monoxide or acetylene) is introduced in the gas phase. An energy source such as plasma or a heated coil is used to transfer energy to the gaseous carbon molecule. It helps to "crack" the molecule into reactive atomic carbon. The carbon diffuses towards the substrate, which is heated. Then, it is coated with a catalyst where it will bind. The CVD nanotube synthesis usually consists of a catalyst preparation step and the nanotube synthesis step. The catalyst is usually prepared by sputtering a transition metal onto a substrate. Then, chemical etching or thermal annealing process can be used to induce catalyst particle nucleation. Thermal annealing process can result in cluster formation on the substrate, from which the nanotubes grow. Typical temperature range for this synthesis route is from 650-900°C. The impurities formed during the synthesis process can be removed using suitable purification methods such as, oxidative treatments in the gaseous and liquid phases, acid treatment, micro filtration, thermal treatment and ultrasound methods. The characteristics of the CNTs produced by CVD method depend on the working conditions, which include the temperature and the pressure of operation, the volume and concentration of gaseous carbon source, the size and pretreatment of metallic catalyst and the time of reaction. Figure 4.26 shows SEM images of aligned carbon nanotubes.

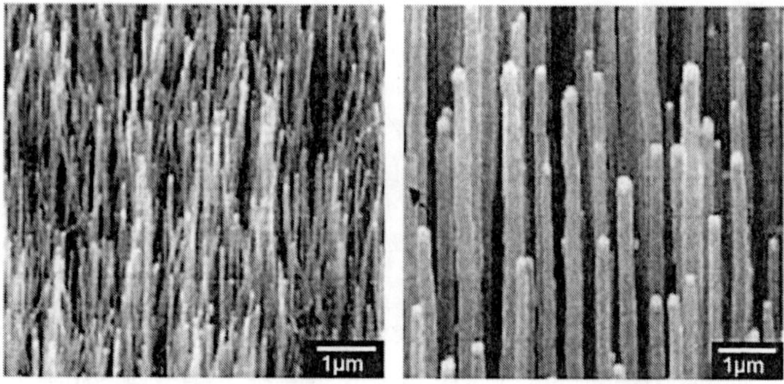

Fig. 4.26: SEM images of aligned carbon nanotubes.

The other techniques for the carbon nanotubes synthesis with CVD are plasma enhanced CVD, thermal chemical CVD, alcohol catalytic CVD, vapour phase growth, aerogel supported CVD and laser assisted CVD.

REFERENCES

1. Dyson, E.W., *Speciality Polymers* (Chapman and Hall: NY, 1987).
2. Mc Cord, M.A. and Rooks, M.J., *SPIE Handbook of Microlithography, Micromachining and Microfabrication* (2000).
3. Martin, C., Rius, G., Llobera, A., Voigt, A., Gruetzner, G. and Perez-Murano, P., *Microelectronic Engg.*, 85 (5-8), 1096-99 (2007).
4. Henrik, Bruus, *Introduction to Nanotechnology* (Lyngby, spring 2004).
5. Richard D.P., Jin, Zhu, Feng, X., Seunghun, H. and Chad, A.M., *Science*, 283(5402), 661-63 (1999).
6. Ruiz, S.A. and Chan, C.S., *Microcontact Printing: A Tool to Pattern*, Soft matter year (2007).
7. Polman, A., et. al., *J. Appl. Phys.*, Vol. 75, No. 6 (1994).
8. Mirabella, S., et. al., *Appl. Phys. Lett.*, 86, 121905 (2005).
9. Rianaldi, F., *Annual Rep.*, Uni. of Ulm (2002).
10. Ploog, K., 27th *Intl Conf. on the Phys. of Semicond.* (2004).
11. Hartman, M.A., and Sitter, H., *Molecular Beam Epitaxy*, Springer, NY (1996).
12. Kuno, M., Introduction to Nanoscience and Nanotechnology: A work book (2005).
13. Gonda, S. and Matsushima, Y., Molecular beam epitaxy of GaP and GaAsP, *Jap. J. Appl. Phys.*, 15, 2093-2101 (1976).

14. Kop'ev, P.S., and Ledentsov, N.N., *Sov. Phys. Semicond.*, 22 10, 1093-1101 (1988).
15. Ritala, M. and Leskela, M., *Nanotechnology,* 10, 19-24 (1999).
16. Suntola, T., *Mater. Sci. Rep.*, 4, 261 (1989).
17. Ritala, M., *Appl. Surf. Sci.*, 112, 223 (1997).
18. Haukka, S., Lakkomaa, E.L. and Root, A., *J. Phys. Chem.*, 97, 5085 (1993).
19. Haukka, S., Lakkomaa, E.L. and Suntola, T., *Thin Solid Films*, 225, 280 (1993).
20. Behrisch, R. (ed.), *Sputtering by particle bombardment* (Springer, Berlin, (1981).
21. Richard, C.J., *Film Deposition: Introduction to Microelectronic Fabrication* (Upper Saddle River: Prentice Hall, 2002).
22. Donald, Smith, *Thin-Film Deposition: Principles and Practice* (MacGraw-Hill, 1995).
23. Dobkin and Zuraw, *Principles of Chemical Vapor Deposition*, Kluwer (2003).
24. Ren et. al., *Appl. Phys. Lett.*, 75, 8 (1999).
25. Ren et. al. *Science*, 282, 5391 (1998).
26. Qingwen, L., Jin, Y. and Zhogfan, L., *Adv. Nanomater Nanodevice*, 59-71 (2002).
27. Yudasaka, et. al., *Appl. Phys. Lett.* 67, 17 (1995).
28. Yudasaka, et. al., *Appl. Phys. Lett.* 70, 14 (1997).
29. Paradise, M. and Goswami, T., *Materials and Design*, 28, 1477-89 (2007).
30. Chen, M., Chen, C.M. and Chen, C.F., *J. Mat. Sci.*, 37, 17, 3561-67 (2002).
31. Gerald, G.S., *Organometallic Vapor-Phase Epitaxy: Theory and Practice* (2nd ed.), (Academic Press, 1999).
32. Manijeh Razeghi, *The MOCVD Challenge: Volumes 1 and 2* (Institute of Physics Publishing, 1995).
33. Brinker, C.J. and Scherer, G.W., *Sol-Gel Science: The Physics and Chemistry of Sol-Gel Processing* (Academic Press, 1990).
34. Hench, L.L. and West, J.K., *Chem. Rev.* 90, 33 (1990).
35. Matijevic, E., *Langmuir*, Vol. 2, p. 12 (1986).
36. Mukherjee, S.P. and Zarzycki J., *J. Am. Ceram. Soc.*, Vol. 62 (1979).
37. Brinker, C.J. and Mukherjee S.P., *J. Mat. Sci.,* Vol. 16, p. 1980 (1981).

38. Gerald, Venzl and John, Ross, *J. Chem, Phys.*, 77 (3), 6 (1982).
39. Reynhout, XEE and Reijenga Ir JC, *The wondrous world of carbon nanotubes* (Eindhoven University of Technology, 2003).
40. Jung, et. al., *Appl. Phys. A—Mat. Sci. and Processing*, 76 (2), 285-86 (2003).
41. Ebbesen, T.W. and Ajayan, P.M., *Nature*, 358, 220-22 (1992).
42. Journet, C. and Bernier, P., *Appl. Phys. A—Mat. Sci. and Processing*, 67 (1), 1-9 (1998).
43. Valentin, N.P. and Philippe, L. (ed.), *Carbon Nanotubes* (NATO Science Series, Springer, 2005).
44. Guo, et. al., *Chem. Phys. Lett.*, 243, (1, 2) (1995).
45. Yudasaka, et. al., *J. Physc. Chem. B* 103, 30 (1999).
46. Eklund, et. al., *Nano letters* 2, 6 (2002).
47. Maser, et. al., *Chem. Phys. Letters*, 292, (4, 5, 6) (1998).
48. Boshakov, et. al., *Diamond and Related Materials*, 11, 3-6 (2002).
49. Scott, C.D., Arepalli, S., Nikolaev, P. and Smalley, R.E., *Appl. Phys. A: Mat. Sci. & Processing*, 72, 5 (2001).
50. Nalwa, H.S. (Ed.), *Handbook of Nanostructured Materials and Nanotechnology* (Academic Press: San Diego, 2000).

5

Tools of Nanomaterials

5.1 Introduction

The world of nanoscale materials is a constant bustle of activity. Nanoscale objects absorb, emit, make and break bond, vibrate and travel; they are always active. It is difficult to catch them and also nanosize objects are too small to be observed with naked eye. Studying these tiny structures require special and deviously clever instruments that measure some properties of matter. Detecting and analyzing the light absorbed or emitted by atoms or molecules is the science of spectroscopy like Infrared, Raman and Ultraviolet-Visible spectroscopy. Energy level transitions, energy and wavelength of electromagnetic spectrum are depicted in the electromagnetic ruler (figure 5.1).

In the mid-1980's a number of high resolution microscopic instruments have been invented, which help the advancement of nanotechnology to a great extent. These microscopic techniques include scanning probe and electron microscopic techniques. The scanning tunnelling microscope and atomic force microscope are examples of scanning probe microscopic techniques whereas transmission electron microscope and scanning electron microscope are examples of electron microscopic technique. The discovery of these instruments has led to numerous inventions on fundamental science. These high resolution instruments have also improved the synthesis of nanostructures by bottom up approach, where the structures are made by assembling atoms and molecules. The intention of this chapter is to provide a basic information about the fundamentals of various structural characterisation tools, such as various spectroscopic and microscopic techniques, that are most widely used in characterising nanomaterials and nanostructures.[1-17]

5.2 X-ray Diffraction (XRD)

XRD is a very good tool for the determination of crystallinity, crystal structures and lattice constants of nanoparticles, nanowires and thin films.[18] This experimental technique is also being used for applications, such as materials identification. XRD is the best method to examine whether a resultant material is amorphous or crystalline in nature. Crystalline phases can be identified by comparing the 'd' values obtained from XRD data with the fundamental data in Joint Committee on Powder Diffraction Standards (JCPDS).

A narrow beam of X-rays from a source is made incident on a specimen at an angle θ with the crystal plane. The wavelength of X-rays usually ranges from 0.7 to 2 Å. The beam is diffracted by the crystalline phases of the specimen. By Bragg's law:

$$2d \sin\theta = n\lambda, \quad (n = 1, 2, 3, \ldots) \qquad \ldots(5.1)$$

where d is the interplanar spacing and λ is the wavelength of the incident X-rays. The intensity of the diffracted beam depends on the diffraction angle 2θ and the specimen's orientation. The study of diffraction pattern helps to measure specimen's structural properties because each sample produces X-rays of definite wavelengths which are characteristic of that material.

XRD is a suitable tool for characterising homogeneous and inhomogeneous strains. From XRD pattern, one can easily measure diffraction peak positions. The diffraction peak positions will be shifted due to homogeneous or uniform elastic strain. The change in interplanar spacing d can be calculated from the shift. The variation of d-spacing is due to the variation of lattice constants under a strain. The diffraction peaks are usually broadened and this broadening is mainly due to inhomogeneous strains. The inhomogeneous strains change from crystals to crystals or within a crystal and this is the cause of peak broadening. The broadening increases with $\sin\theta$. Another reason for peak broadening is the finite size of crystals and in such cases, the broadening is independent of $\sin\theta$. Careful analysis of peak shapes is necessary to find the contribution of both inhomogeneous strain and crystallite size to the peak width.

The size D of the crystallite can be calculated from the width of peak using Scherrer's formula, provided there is no inhomogeneous strain.

$$D = K\lambda/(B \cos\theta_B) \quad \text{...(5.2)}$$

where K is Scherrer's constant, λ is the wavelength of X-rays, B is the full width at half maximum (FWHM) of the diffraction peak and θ_B is the diffraction angle. XRD of low Z elements are less sensitive as compared to high Z elements because the intensity of X-rays diffracted from low Z elements is low. XRD commonly requires large specimens because of small diffraction intensities. The information acquired in XRD is an average over a large quantity of materials. This technique is very useful for nanoparticle characterisation because it works only at very small dimensions. XRD can also be used to estimate the thickness of thin films.

5.3 Infrared (IR) Spectroscopy

IR spectroscopy is an important tool for sample identification and structural elucidation. This spectroscopic technique measures absorption of various IR frequencies by a specimen.[5] IR spectroscopy with suitable sampling accessories helps us to use solid, liquid and gas samples.

Each atom in a molecule vibrates as the temperature is above absolute zero. When the frequency of vibration and the frequency of the incident IR beam are equal, the molecule absorbs the radiation. The method of analyzing this vibration is called infrared spectroscopy. Each atom has 3 degrees of freedom, corresponds to motion along any of the three coordinate axes (x, y, z). If a polyatomic molecule consists of n atoms, each atom has 3n degrees of freedom. Three of the degrees of freedom are used to describe translatory motion and another 3 degrees of freedom are used to describe the rotation of the molecule. The remaining (3n – 6) degrees of freedom are fundamental vibrations for non-linear molecules. However, linear molecules have (3n – 5) fundamental modes of vibration since they require only 2 degrees of freedom to describe rotation. Only those fundamental modes of vibration, that produce a net change in the dipole moment can generate IR activity. The fundamental modes which causes polarizability changes can result in Raman activity. Obviously, some vibrations are both IR and Raman active.

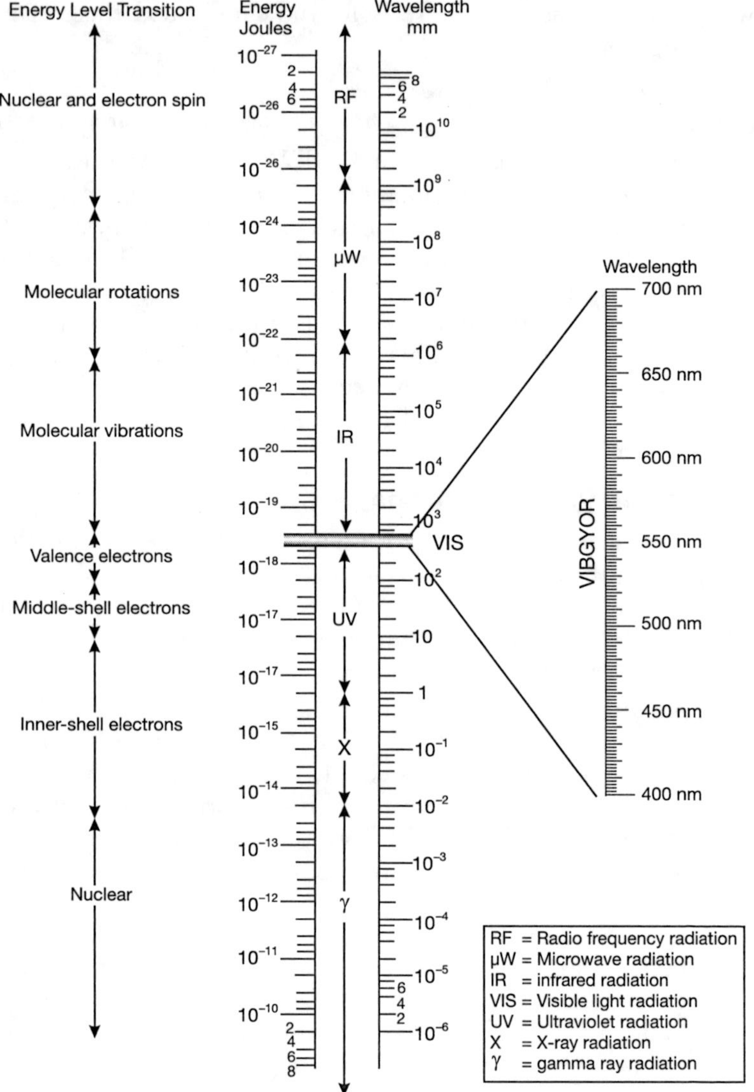

Fig. 5.1: Electromagnetic radiation ruler.

Source: www.upload.wikimedia.org

When IR radiation is incident on a sample, some frequencies are absorbed by the sample and some other frequencies are transmitted. The resulting spectrum consists of absorption peaks, corresponding to the frequencies of vibrations between the bonds of the atoms. It is significant that IR spectrum of different materials is different since each material is a unique combination of atoms. Clearly, IR spectroscopy gives a qualitative analysis of materials. Apart from this, IR spectroscopic technique gives a measure of the amount of materials present in the sample because the size of the absorption peak is a direct indication of quantitative analysis of materials. The modern software based IR spectroscopy works as an excellent tool for quantitative analysis of materials. Figure 5.2 shows the sketch of a typical IR spectrometer.

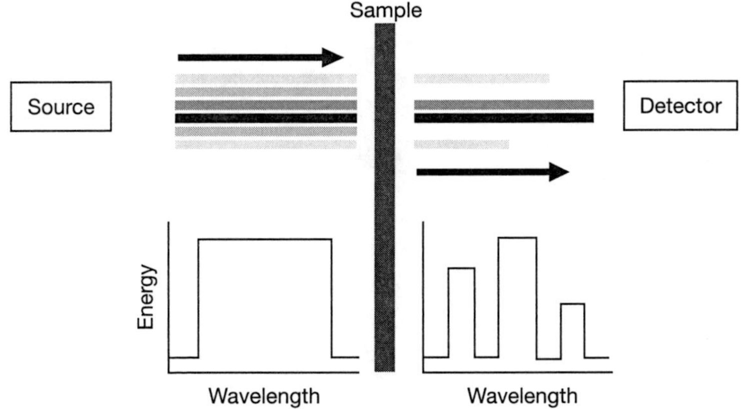

Fig. 5.2: Schematic representation of IR spectroscope.

Radiation source, monochromator and detector are the main components of IR spectrometer. Generally, an electrically heated inert solid (1000-1800°C) is used as the radiation source. The monochromator disperse the wide spectrum of radiation and offers IR radiation of suitable frequency range. For this, a combination of prisms or gratings with variable slit mechanisms, mirrors and filters is used. Photon detectors and thermal detectors, such as thermocouples, thermistors and Golay detectors are the two usual types of IR detectors used in IR spectrometer. Thermal detectors measure the heating effects generated by IR radiation. For example, thermistor measures electrical resistance and thermocouple measures voltage at a junction of dissimilar metals. Furthermore, the thermal detectors exhibit a linear response even for a wide range of frequencies. However, they are not as sensitive and fast as photon detectors. The working of the photon detectors are based on interaction of IR radiation with a semiconductor material. As a result of the interaction, electrons from the valence band are excited to conduction band and hence, a small current or voltage is developed.

5.3.1 Fourier Transform Infrared (FTIR) Spectroscopy

Fourier transform spectrometers have superior speed and sensitivity as compared to dispersive instruments for most applications. In FTIR spectroscopy, all component frequencies are viewed simultaneously.[6] All molecules absorb IR light except monoatomic (e.g., He, Ne, Ar) and homopolar diatomic (e.g., H_2, N_2, O_2) molecules. Molecules only absorb some IR radiations, whose frequencies affect the dipole moment of the molecule. The dipole moment of a molecule is due to the differences of charges in the electronic fields of its atoms. Molecules having only a dipole moment interact with infrared photons. The interaction can result in the excitation of molecules to higher vibrational states. The electronic fields of atoms of homopolar diatomic molecules are equal and hence, they do not poses a dipole moment. No dipole moment is associated with monoatomic molecules also because they have only one atom. Since monoatomic and homopolar diatomic molecules have no dipole moments, they do not absorb IR radiation.

As discussed above, each molecule absorbs only IR light of certain frequencies based on the characteristics of each molecule. Hence, the study of the absorption spectrum helps to identify the type of molecule (qualitative analysis) and the amount of molecule present in the sample (quantitative analysis).

The radiation source, interferometer and detector are the main components of an FTIR system. Figure 5.3 illustrates the working of a typical FTIR spectrometer. The IR radiation source of FTIR spectrometer contains a water cooling system to ensure better power and stability. An interferometer is used instead of monochromator, which divides incident beams and creates an optical path difference between the beams. The superposition of these beams produce interference pattern, which contain infrared spectral information. The interferometer used in FTIR spectrometer is the Michelson interferometer, which consists of a fixed mirror, a moving mirror and a beamsplitter. The two mirrors are mounted at right angles to each other. A semi-reflecting device is used as the beamsplitter. The beamsplitter splits the incident beam and transmits half of the beam to the fixed mirror and the rest is reflected to the moving mirror. The beams after reflection from the mirrors are superposed at the beamsplitter and an interference pattern is generated. The resulting beam is then passed through the sample, and finally collected by the detector.

If the velocity of the moving mirror is constant, the intensity of beam collecting the detector changes sinusoidally. The resulting interferogram output is shown in figure 5.3. The interferogram is a time domain spectrum that records the detector response changes versus time within the mirror scan. When the sample absorbs frequency of radiation, the amplitude of the sinusoidal

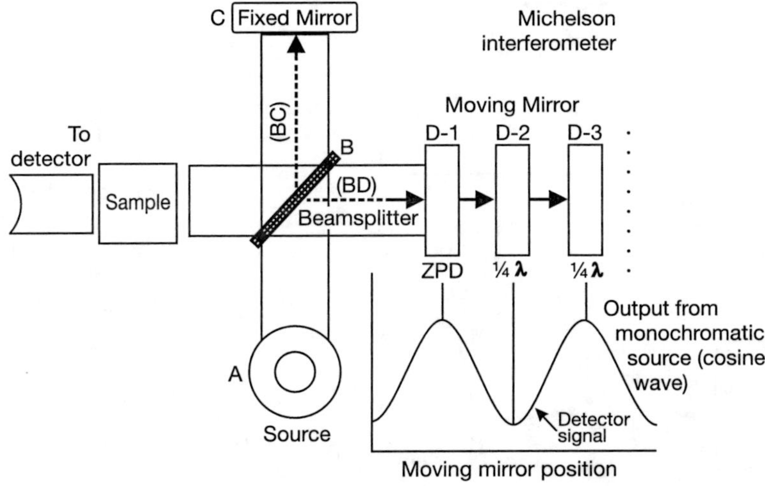

Fig. 5.3: Schematic diagram of a typical FTIR spectrometer.

wave will be reduced. Fourier transformation converts the time domain interferogram to frequency domain spectrum showing intensity versus frequency, the final IR spectrum. The FTIR spectroscopy has many advantages over dispersive method of infrared spectral analysis. This non-destructive and fast technique gives precise measurement, which requires no external calibration.

5.4 Raman Spectroscopy

This technique is based on inelastic scattering of monochromatic light. In inelastic scattering, the frequency of light changes due to its interaction with a sample. Laser beams from the source are absorbed by the sample and re-emitted during de-excitation. The frequency of the emerging beam consists of both lower and higher frequencies in addition to the original frequency. This phenomenon is called the Raman Effect. The study of the change in frequencies gives information about transitions (vibrational, rotational and other low frequency) in molecules of solid, liquid or gaseous samples.[7-8]

A typical Raman spectrometer consists of a source of light such as a laser, a double or triple monochromator and a signal processing unit composed of a detector, an amplifier and an output device. A sketch of a typical Raman spectrometer is shown in figure 5.4.

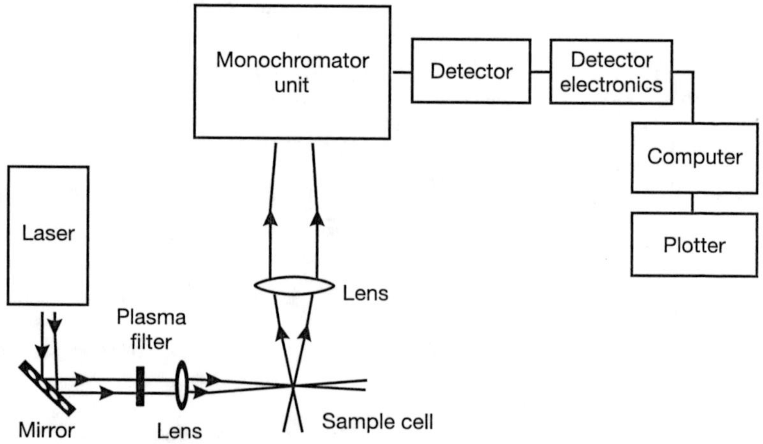

Fig. 5.4: Schematic diagram of a laser Raman system.
Source: www.emeraldinsight.com

A number of stages are involved in the acquisition of Raman spectrum. The sample on the sample chamber is illuminated with a laser light. A convex lens is used to focus light on the sample. Generally, liquid and solid samples are taken in a Pyrex capillary tube. The scattered light is collected and focused at the entrance slit of the monochromator using another lens. The width of the slit is adjusted for creating suitable spectral resolution. The light coming out through the exit slit of the monochromator is collected and focused on the surface of a detector. The detector converts this optical signal into an electrical signal and is further processed using detector electronics. Computer stores the output signal from the detector electronics for each predetermined frequency interval. A plot of the signal intensity versus wave number constitutes a Raman spectrum. The advances in solid state detectors have replaced the conventional photomultiplier tube detectors with multichannel detectors (MCD). MCD based laser Raman spectrometers are more efficient and faster than the conventional Raman systems.

5.5 UV-Vis Spectroscopy

UV-Vis spectroscopic technique is useful to characterise the absorption, transmission, and reflectivity of materials.[9] The principle of UV-Vis spectrometer is based on the fact that molecules of the sample absorb UV or visible light and results in the excitation of outer electrons in the molecule. This absorption spectroscopy measures the absorption of light after it interacts with the sample. The measurements can be done either at a monochromatic wavelength or a wide spectral range. The wavelength of light absorbed by the sample is the characteristic of its chemical structure. Different spectral regions of the UV-Vis spectrum are absorbed by different types of molecules. As figure 5.1 shows, absorption of microwave radiation can result in molecular rotational motion, and IR absorption results in vibrational motion of molecules. But UV-Vis absorption can result transition of electrons to higher energy states. Each molecule undergoes electronic excitation following absorption of light. UV-visible light is adequate for molecules having conjugated electron systems (e.g., benzene absorbs light of wavelength 260 nm).

The absorption spectrum shifts to the lower energy as the degree of conjugation increases (e.g., naphthalene absorbs light up to 300 nm and anthracene up to 400 nm). The study of the absorption spectra helps to identify atomic and molecular species, since they are characteristic of molecular structure.

The relation connecting the intensity of transmitted light through a solution of an absorbing material and the concentration of the material is given by Beer-Lambert law. According to the law:

$$-\log (I/I_0) = A = \varepsilon_\lambda bc, \qquad ...(5.3)$$

where I_0 and I are the intensity of incident and transmitted light respectively, A the absorbance, ε_λ the molar absorptivity (litre/mole/cm), b the cell path length (cm) and c the concentration of solution (moles/litre). The molar absorptivity represents the spectrum of the solution and is a function of wavelength. The value ε_λ is represented for a particular wavelength (e.g. ε_{532}). Thus, the study of UV-Vis spectroscopy helps to measure the quantity of chemicals present in a solution.

UV-Vis Spectrophotometer consists of a UV light source, a monochromator and a detector. The monochromator works as a diffraction grating to dispense the beam of light into the various wavelengths. The detectors role is to record the intensity of light which has been transmitted. Before the samples are run, a reference must be taken first to calibrate the spectra to screen out any spectral interference. In the case of liquid samples, the solvent which has been used to dissolve the sample is used. However, there is a criterion that the solvent should not absorb UV radiation in the same region as the sample being analysed. In the case of solid state UV-Vis spectroscopy, the reference is normally KBr as it does not absorb radiation in the same region as most samples. This method is often used with samples that IR spectroscopy fails to identify.

5.6 Scanning Probe Microscopy (SPM)

SPM technique scans sample surfaces using a probe tip. The SPM has a great impact on many areas of science and technology. This technique is highly useful for the investigation and manipulation of nanoscale materials. The probe of SPM is a narrow tip having a radius of curvature around 3-50 nm. The tip of the scanning probe is fixed on a flexible cantilever, which allows the probe tip to scan the surface profile. As the probe tip moves within the proximity of the sample surface, the interactive forces between the probe tip and surface affect the movement of the cantilever. The movements of the cantilever can be detected using suitable sensors. Various types of interactions can be studied with the help of SPM, depending on the mechanics of the probe. The SPM techniques can help to develop image clusters of individual atoms and molecules because they can operate even upto nanometers.[1-3]

Based on interactions, SPMs are of various types such as, scanning tunnelling microscopy (STM), atomic force microscopy (AFM) and near-field scanning optical microscopy (NSOM). STM measures electronic tunnelling current, AFM measures interaction forces and NSOM measures local optical properties using near field effects (figure 5.5). These characterisation tools help to study the structural, mechanical, electrical and optical properties of materials in any environment.[17]

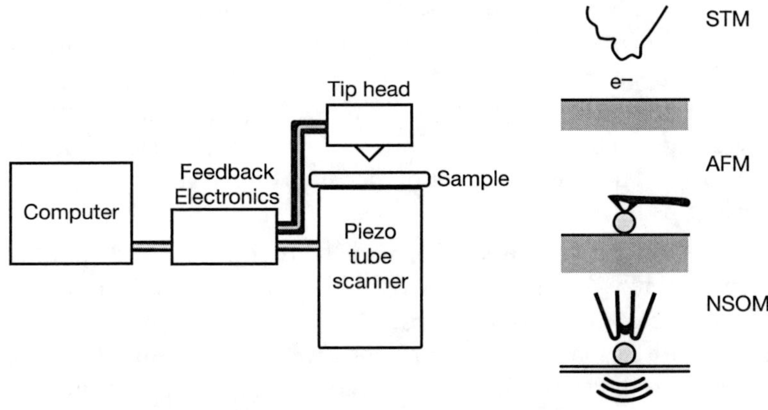

Fig. 5.5: Schematic diagram of (a) a typical SPM and (b) the signals observed in STM, AFM and NSOM techniques.[2]

5.6.1 Scanning Tunnelling Microscopy (STM)

In 1981, Gerd K. Binnig and Heinrich Rohrer invented STM. They were awarded half of the 1986 Nobel Laureate for their work in Physics. Three-dimensional image of solid surface with atomic resolution was first obtained with STM.[10, 11] However, STMs can only be used to study electrically conducting samples. Mainly there are five scientific processes that the STM integrates to make atomic resolution. They are: (i) quantum mechanical tunnelling, (ii) controlled motion over small distances by piezoelectrics, (iii) negative feedback, (iv) vibration isolation, and (v) electronic data collection. Each of these processes was known to the scientific community even before the invention of STM.

The electron tunnelling principle was introduced by Giaever. Consider a system of two metals separated by an insulating thin film. When a suitable potential difference is applied across the metals, a current will start flowing because electrons can penetrate the potential barrier. It is possible to measure the tunnelling current by spacing the two metals less than 10 nm. Vacuum tunnelling together with lateral scanning was introduced by Binnig and his co-workers. The vacuum offers the ideal barrier for tunnelling. The lateral scanning can help to image sample surfaces with wonderful resolution, even sufficient to image single atoms. Since the tunnelling current varies exponentially with the distance between the metal tip and the scanned surface, vertical resolution of the STM is very high. Usually, tunnelling current reduces as the separation between the electrodes is increased. The lateral resolution depends on the sharpness of the probe tips.

The tip of a typical STM is made from tungsten or Pt-Ir alloy that is attached to a piezo-drive. Three mutually perpendicular piezoelectric transducers: x-piezo, y-piezo and z-piezo are available with the piezo-drive, as shown in figure 5.6. When a p.d. is applied, piezoelectric transducer expands or contracts. The scanning waveform is formed between the x – y piezos and make the tip raster scan over the sample surface. The separation between the probe tip and the surface of the sample is kept very narrow (a fraction of nanometre) using the coarse positioner and the z-piezo. A suitable potential difference (0.1 V – 1 V) is applied across the tip and the sample. It can induce a tunnelling current of about 0.1 nA to 1 nA. A negative feedback mechanism is employed to control the z-piezo, and to maintain a constant tunnelling current. The voltage level on the z-piezo indicates the local height of the topography.

When the bias potential V between the tip and the sample is greater than zero (V > 0), electrons tunnel from the tip into the sample surface. Though if V < 0, the electrons from the sample surface tunnel into the tip. By using a current amplifier, the tunnelling current can be converted into a voltage. Compare this voltage with a reference value. The difference voltage is amplified to drive the z-piezo. The phase of the amplifier is chosen to provide a negative feedback. If tunnelling current is higher than the reference value, the voltage applied to the z-piezo tends to withdraw the tip from the surface, and *vice versa*. Therefore an equilibrium z position is developed. As the tip scans over the xy plane, a two-dimensional array of equilibrium z positions will be obtained. This equilibrium position of the z-piezo represents a contour plot of the equal tunnelling current. The contour plot can be displayed and stored in the computer.

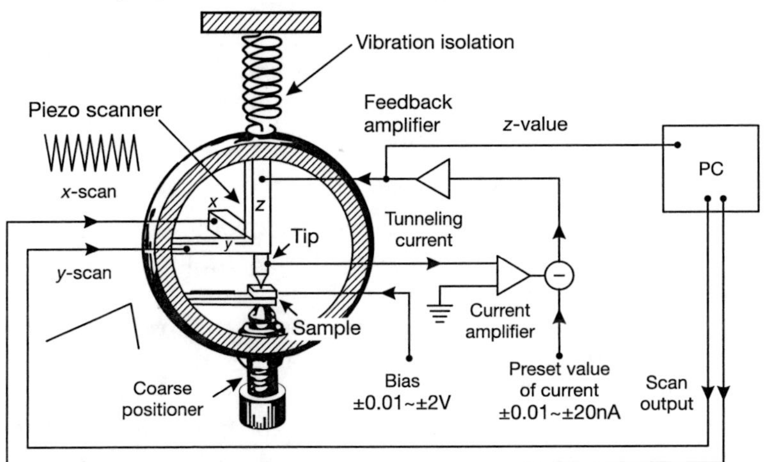

Fig. 5.6: The schematic representation of a typical STM.

The topography of the surface is shown on a display monitor as shown in figure 5.7(a). In the figure, the bright and dark spots represent high z and low z values respectively. The scale bar represents the z values corresponding to the grey levels.

Figure 5.7(b) shows contour plot along a given line, a quantitative representation of the topography. The unit for x and y is nanometre and that for z is picometer (10^{-12} m). The STM unit is made as rigid as possible to ensure vibration isolation, and thereby achieve atomic resolution.

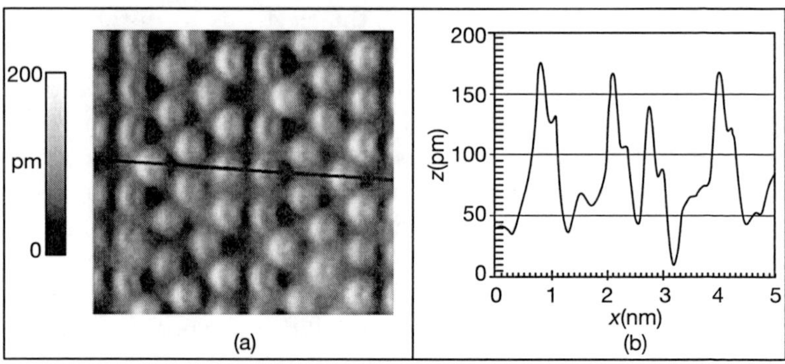

Fig. 5.7: Model grey scale image and contour plot.

One disadvantage of the STM is that it requires conductive samples, so that at least some features of the specimen must show electrical conductivity to some extend. To circumvent this limitation, AFM was subsequently developed and it is used for studies of non-conducting samples.

5.6.2 Atomic Force Microscopy (AFM)

In 1986, Binnig and Gerber discovered atomic force microscope to measure ultra small forces (< 1 μN), which is present between the probe tip and the sample surface. This microscopic technique is an important tool for nanoscale imaging. It also helps to manipulate nanoscale materials. Besides these, AFM technique helps to study some physical and chemical properties of samples. One significant advantage is that AFM technique is suitable for studies of both electrically conducting and insulating samples.[3] The resolution of the AFM is very high, of the order of fractions of a nanometre. The force of interaction between the tip and the sample surface depends on the separation between them and also nature of the sample surface.

Basic Principles

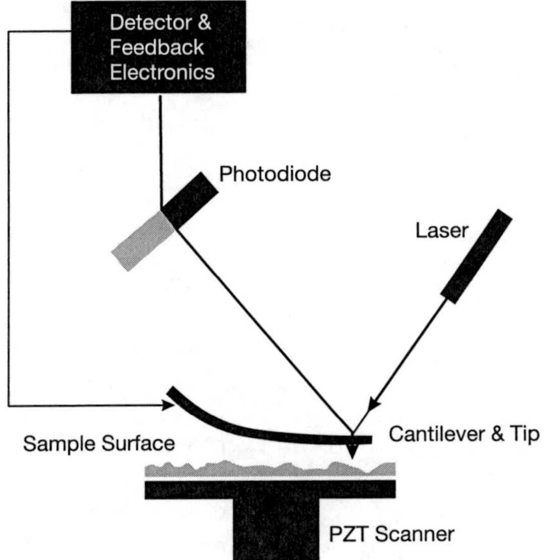

Fig. 5.8: Schematic sketch of a typical AFM.

Source: www.iap.tuwien.ac.at

A typical AFM consists of a cantilever with a probe tip, laser beam deflection system, a photodiode, PZT scanner, electronic feed back system and a detector (figure 5.8). The sharp tip attached to the end of the cantilever scans the surface of the specimen. The cantilever is made from Si or SiN. Its spring constant is about 0.01-100 N/m, and resonant frequency is about 5-300 kHz. When the tip of the probe moves near the surface of the specimen, the interaction force will cause a deflection of the cantilever. By Hooke's law, Force F = –kx, where k is the spring constant and x is the distance of the probe from the sample. Force distance curve is shown in figure 5.9.

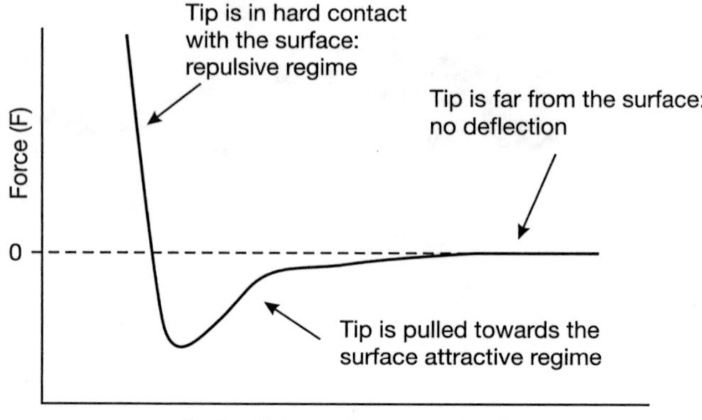

Fig. 5.9: Force distance curve of an AFM.

The deflection of the cantilever is measured using a laser beam deflection technique, as shown in figure 5.10. Beam of light from the laser source is reflected from the top surface of the cantilever. The reflected beam is made to incident on a split photodiode. Cantilever deflections are proportional to the difference signal $V_A - V_B$. It is possible to measure even sub-angstrom deflections. Hence, very small forces about tens of pico-Newtons can be measured. Beside this laser deflection method, methods like optical interferometry, capacitive sensing and piezoresistive cantilevers are used for the detection of cantilever deflection. However, these methods are not as sensitive as laser deflection method.

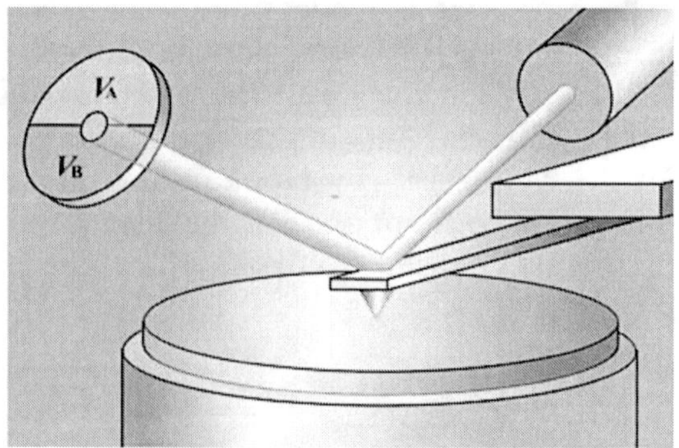

Fig. 5.10: Optical beam deflection system that detects cantilever motion in the AFM.[2]

If the tip scans at a constant height, it collides with the sample surface and will be damaged. To overcome this difficulty, an electronic feedback system is used to adjust the tip-to-sample distance to maintain a steady force between them. The specimen to be scanned is placed on a piezoelectric tube. The tube moves the specimen along the z-axis for keeping a steady force. It also moves the sample along the x and y axes for scanning the surface of the specimen. Alternatively a tripod configuration of three piezo crystals can be used for scanning, as discussed in section 5.6.1. The advantage of this method is that it eliminates the distortion effects caused by a tube scanner. However in new systems, the probe tip is mounted on a vertical piezo scanner and the sample to be scanned is placed on a new piezo block. The resulting plot of the area s = f(x, y) represents the topography of the sample.

Imaging Modes

It is possible to operate AFM in different imaging modes, depending on the application or requirement. Static or contact mode and dynamic or non-contact mode are the two basic modes of operation of AFM. In the first mode, the static tip deflection signal will be used as a feedback signal. Generally, a low stiffness cantilever is employed to enhance the deflection signal. But, within the proximity of the sample surface, attractive forces are strong as to cause the tip to collide the surface (figure 5.9). In order to maintain a steady deflection, the interaction force is kept steady during scanning.

In the non-contact mode, there is no direct contact between the tip of the cantilever and the sample surface. However, the cantilever oscillates at a frequency slightly greater than its resonance frequency. The amplitude of oscillation of the cantilever is typically a few nanometres. The characteristics of oscillation, such as amplitude, phase and resonance frequency are modified by tip-sample interaction forces. These modifications in oscillation with respect to the external reference oscillation give information about the characteristics of the specimen. Both the frequency modulation (FM) and amplitude modulation (AM) schemes are used in non-contact mode operation. In FM scheme, the frequency of oscillation will be modified and may provide information about tip-sample interactions. True atomic resolution will be obtained at ultra-high vacuum conditions. In AM, the amplitude or phase of oscillation will be modified and may provide necessary feedback signal for imaging. The modifications in the phase can help to explore various materials present on the sample surface. If the tip used in an AFM is sharper, the better the resolution. Carbon nanotubes are best suited for AFM tips because they are very strong and flexible, and also it may not be broken even though much force is exerted on it.

Besides imaging, force spectroscopy is another major application of AFM. Force spectroscopy helps to study and measure nanoscale contacts, atomic bonding, Casimir forces and van der Waals forces.

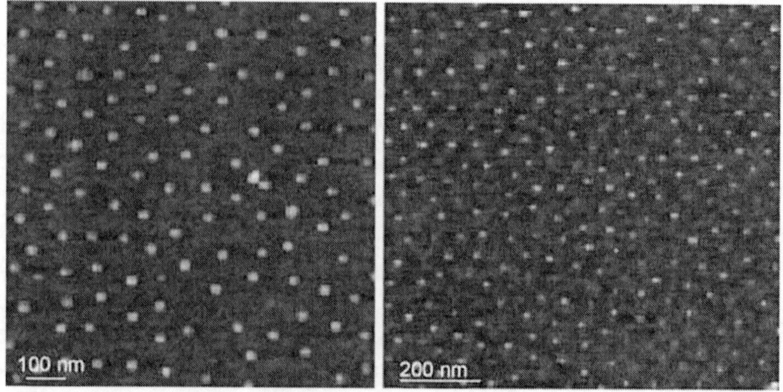

Fig. 5.11: AFM images of Iridium (left) and Gold (right) nanoparticles.

Source: www.physics.ucf.edu

5.6.3 Near-Field Scanning Optical Microscopy (NSOM)

In 1920, Synge introduced near-field scanning optical microscopic technique with remarkable accuracy. This technique provides the highest lateral optical resolution. NSOM scans a small aperture (~100 nm) located very close to the specimen and detection of this light energy forms the image of the surface. In the illumination mode of NSOM, a dielectric probe positioned at a distance $d \ll \lambda$ from the surface illuminates the sample from above. Either the reflected or the transmitted beam of light is collected in the far field using detectors. A dielectric probe in the near field collects the transmitted light. In a typical NSOM, a tapered single-mode optical fibre tip scans over the sample surface using piezoelectric transducers. The tip is coated with a very narrow film of metals to reduce loss of light. This technique helps to obtain topographic and light intensity images simultaneously. NSOM optical fibre tip scans the surface of the sample, collects the total reflected intensity from each scan point and integrates it as one pixel to form the final image, as shown in figure 5.12. Figure 5.13 compares the resolution capabilities of scanning electron microscopy, conventional optical microscopy and NSOM.

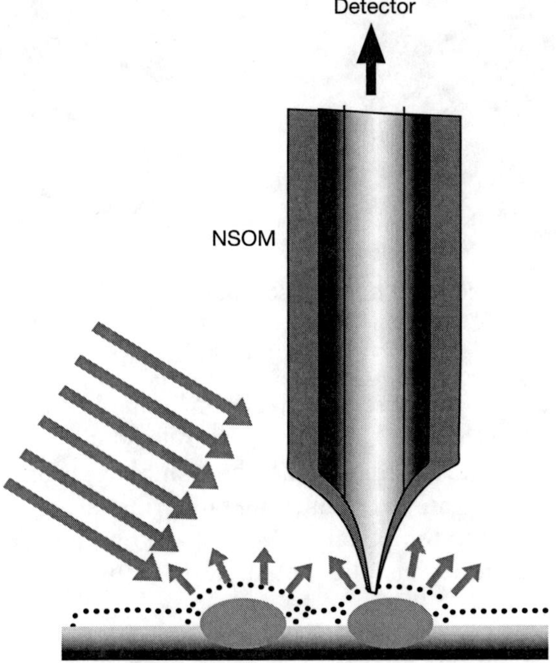

Fig. 5.12: Schematic representation of NSOM.

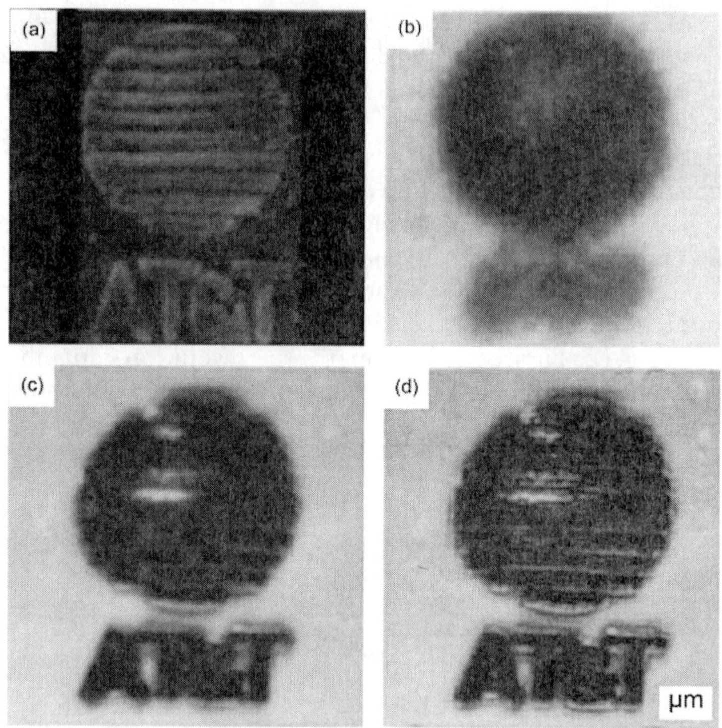

Fig. 5.13: Resolution comparison of (a) SEM, (b) conventional optical microscopy, (c) NSOM image, and (d) NSOM image after Fourier deconvolution.

Besides optical fibre based NSOM, there is another type called apertureless NSOM. The apertureless NSOM is based on a sharp metallic tip that scans the sample. Technically it is possible to fabricate very sharp metallic tips (atomic size STM tips) as compared to optical fibre tips (~50 nm diameter). Thus, NSOM technique has the potential to achieve much better resolution.

5.7 Electron Microscopies

5.7.1 Transmission Electron Microscopy (TEM)

TEM is a high resolution microscopic technique capable of imaging fine details even up to a single column of atoms. Its resolution is much larger than that of the light microscopy. This high resolution of TEM is due to the small de Broglie wavelength of electrons.[12, 13] TEMs are widely used in cancer research, virology, pollution, materials science, semiconductor nanoresearch and so on.

A typical TEM consists of a vacuum system, an electron gun which generates electron beam, a number of electromagnetic lenses and electrostatic plates. The electromagnetic lenses and electrostatic plates guide the electron beam as required. Imaging devices are subsequently used to create images which are formed from the interaction of the electrons transmitted through the ultra thin specimen. These images are magnified and focused on a fluorescent screen or a charge coupled device (CCD) camera.

Electrons

In 1923, de Broglie introduced the concept of matter waves, which states that all particles have an associated wavelength linked to their momentum. The de Broglie wavelength

$$\lambda = [h/mv] \quad ...(5.4)$$

where m is the relativistic mass, v the relativistic velocity and h the Planck's constant. This equation is called de Broglie wave equation. Hans Bush (1927) found that it is possible to focus an electron beam using a magnetic coil as a glass lens focus light. In 1931, Ernst Ruska and Max Knoll recorded the first TEM image. Reinhold Rudenberg of Siemens Company took the patent of an electrostatic lens electron microscope during the same year.

The wavelength of electrons (λ_e) can be measured by equating the de Broglie wave equation to the kinetic energy of an electron. A relativistic correction is applied because the velocity of electron approaches the speed of light c.

$$\text{The wavelength, } \lambda_e \sim [h / \sqrt{\{2m_0E\ (1 + E/2m_0c^2)\}}] \quad ...(5.5)$$

The emitted electrons from the gun are accelerated by an electric potential of about 100–1000 kV and their velocity approaches the speed of light (0.6–0.9c). These high speed electrons are focused on the sample surface using electrostatic and electromagnetic lenses. The transmitted electron beam is used to develop images. The magnetic lens aberrations limit the TEM resolution to the Å order. This resolution of the TEM is suitable for material imaging and structure determination at the atomic level.

Source Formation

The electron gun in a TEM generates electrons by thermionic or field electron emission. Schematic layout of optical components in a typical TEM is shown in figure 5.14. The manipulation and focusing of the electron beam is made using electromagnetic and electrostatic effects. When electron enters the magnetic field of electromagnets, they start moving according to right hand rule. It is possible to form electromagnetic lens of variable focusing power. The electrostatic fields can deflect the electrons through desired angle. These two effects together with an electron imaging system can help to control the electron beam path. Furthermore, it is possible to modify the optical configuration of a TEM.

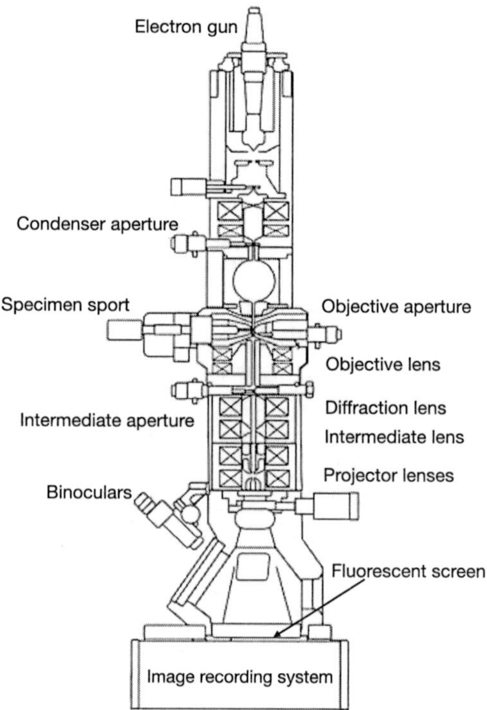

Fig. 5.14: Typical layout of a TEM.

Source: www.en.wikipedia.org

Optics

A typical TEM mainly consists of condensor lenses, objective lenses and projector lenses. The condensor lenses form primary beam, and the beam is focused on the surface of the sample using objective lenses. With the help of projector lenses, the beam is expanded onto the fluorescent screen.

Display

The display unit consists of a phosphor screen and an image recording system. The phosphor screen is made of fine zinc sulphide particles (10-100 nm) for direct observation. The image recording system is a film or doped YAG screen coupled CCDs. It is possible to rotate a sample by a desired angle, and thus to take different images of the sample at different angles (in 10 increments). These images can be used to construct a 3D image of the sample. TEM images of CdSe quantum dots are shown in figure 5.15.

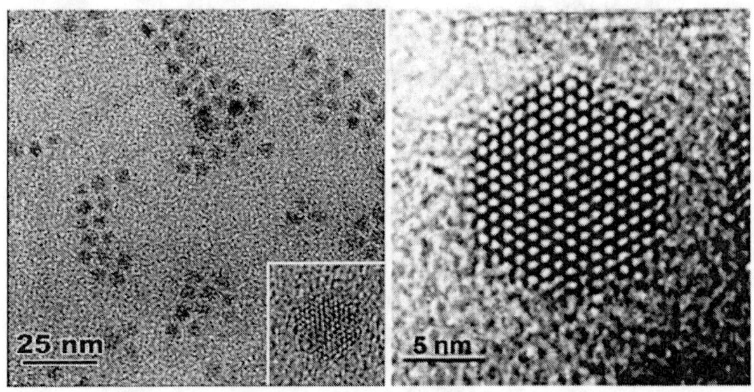

Fig. 5.15: Transmission electron micrographs of CdSe quantum dots.

Source: www.earthlink.net

5.7.2 High-Resolution Transmission Electron Microscopy (HRTEM)

HRTEM is a powerful tool for studying the structure of particles, interfaces and crystal defects. This microscopic technique images crystallographic structure of materials with atomic scale precision.[14] HRTEM provides highest resolution about 0.08 nm with microscopes. Because of this high resolution, this unique tool helps to study the properties of nanocrystalline materials like semiconductors and metals. The phase-contrast imaging is the basis of image formation in HRTEM. In this type of imaging, the contrast may not be interpretable spontaneously because the image is affected by aberrations of imaging lenses.

Basic Principle

A thin sample of crystal is mounted such that a low index direction is at right angles to the incident electron beam direction. The primary beam of electrons will be diffracted by all lattice planes parallel to the electron beam. The diffracted and the primary beam are made incident on the objective lens. Their interference result a back transformation and produces an enlarged image of the periodic potential. This image can be further magnified by the electron optical system and will be formed on the screen. This imaging process is called phase-contrast imaging and the image formed is an indirect depiction of the crystallographic structure of the specimen. The phase of the electron wave carries information about the structure of samples and generates contrast in the image. Hence, the name phase-contrast imaging. The principle of working is schematically illustrated in figure 5.16.

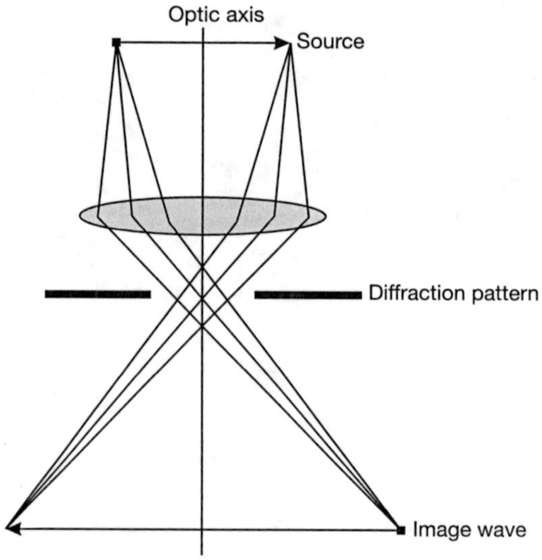

Fig. 5.16: Schematic diagram of the electron beam column in HRTEM.

The interaction of the electron wave with the crystallographic structure of the sample gives a qualitative information about the structure of the specimen. Each imaging electron interacts independently with the sample. The electron wave can be considered as a plane wave and is made incident on the sample surface. When it enters the specimen, the electron beam is attracted by the positive potentials of the atomic cores, and directs it along the atomic columns of the crystallographic lattice. The interaction between the electron waves in different atomic columns results Bragg's diffraction. HRTEM image of a CdSe nanocrystal is shown in figure 5.17.

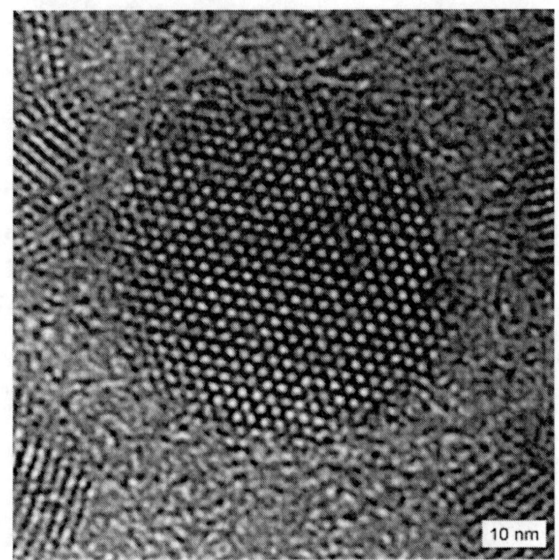

Fig. 5.17: HRTEM image of a CdSe nanocrystal.

Source: www.chemistry.manchester.ac.uk

5.7.3 Scanning Electron Microscopy (SEM)

SEM is widely used for the characterisation of nanomaterials and nanostructures.[15] Its resolution is nearly a few nanometres. The magnification of SEM can be adjusted from 10 to around 300,000. It is interesting that SEM provides the chemical composition information near the surface of the sample in addition to topographical information. High energy electron beam from an electron gun is focused onto the surface of the specimen to produce a number of signals, as shown in figure 5.18. The signals that are derived from the interactions between electron and sample could provide information about the sample, such as chemical composition, crystalline structure and external morphology (texture).

Schematic diagram of a scanning electron microscope is shown in figure 5.18. A SEM is composed of a source of electron, electromagnetic lenses, vacuum system, detectors and a display device. In a SEM, a beam of electrons having energy ranging from a few hundred eV to 50 KeV is made incident on the sample surface. A variety of interactions occur that results in the emission of a number of electrons and photons from the sample. The emitted electrons consist of secondary electrons, backscattered electrons and diffracted backscattered electrons. The SEM employs secondary electrons and backscattered electrons for imaging samples. Surface morphology and topography of the sample can be obtained from the secondary electrons. Crystal structures and orientation of materials will be derived from the diffracted backscattered electrons. The photons emitted during the interaction may consist of characteristic X-rays, continuum X-rays and visible light. When a beam of electrons is incident on the specimen, electrons in discrete orbitals will be excited. During the process of de-excitation, the excited electrons are returned to lower energy states and as a result, characteristic X-rays of fixed wavelength will be produced. These X-rays are suitable for elemental analysis. The scanning electron microscopic technique is a non-destructive method because the X-rays produced during the interaction do not result in any volume loss of the specimen. Hence, the same materials can be analysed repeatedly.

The resolving power, R, of an instrument is defined as:

$$R = \lambda/(2NA) \qquad ...(5.6)$$

where λ is the wavelength of electrons used and NA is the numerical aperture, which is engraved on each objective and condenser lens system, and a measure of the electron gathering ability of the objective, or the electron providing ability of the condenser.

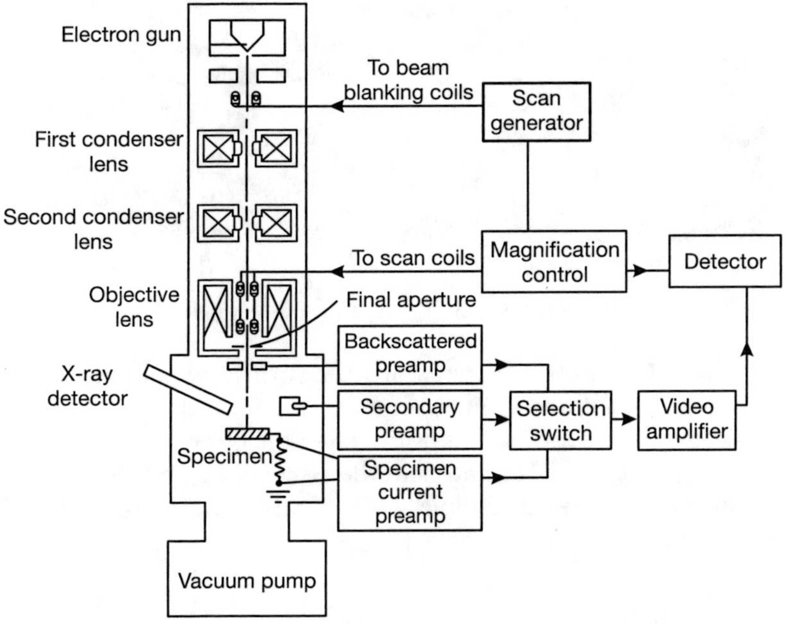

Fig. 5.18: Schematic diagram of a scanning electron microscope.

The electron gun emits a narrow beam of electrons and these electrons are focused using magnetic lenses. A coil of copper wire produces a magnetic field that is shaped into a suitable geometry to create the electromagnetic lensing action, similar to that of an optical lens. If an electron of charge q moves with a velocity v through a magnetic field B, it experiences a radial force F inward, called magnetic Lorenz force. Magnetic Lorenz force $F = q(v \times B)$. The focal length of the electromagnetic lens depends on the gun voltage and the amount of current through the coil. The velocity of the electron beam depends on gun voltage and the flux density depends on the current through the coil. Therefore, it is possible to control the focal length of the electromagnetic lens by controlling the current through the coil. As current increases, the radial force experienced by the beam increases and hence, the focal length of the electromagnetic lens will be reduced.

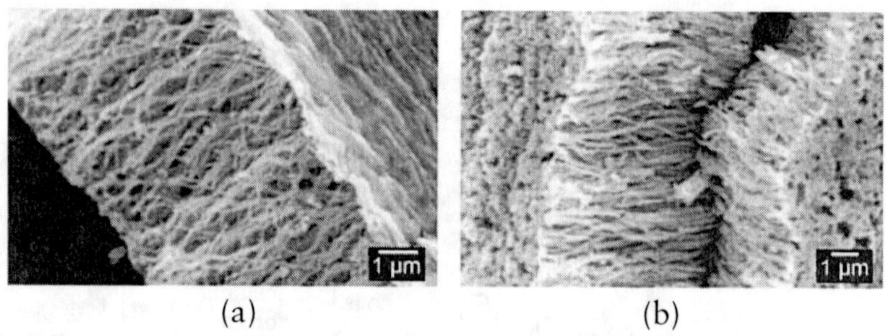

Fig. 5.19: SEM images of nanorod arrays.

In SEM, electron beam scans the surface of sample. There are two coils for scanning, one for raster and the other for deflection. These coils are positioned in the bore of the objective lens cage (figure 5.18) and help the electron beam to scan over a square area on the sample surface. SEM images of nanorod arrays are shown in figure 5.19.

REFERENCES

1. Henrik Bruus, *Introduction to Nanotechnology* (Lyngby, Spring, 2004).
2. Bharath Bhushan, *Handbook of Nanotechnology* (Springer: NY, 2004).
3. Wiesendanger, R., *Scanning Probe Microscopy and Spectroscopy: Methods and Applications* (Cambridge University Press, 1994).
4. James, D. Ingle, Jr. and Stanley, R. Crouch, *Spectrochemical Analysis* (Prentice Hall, 1988).
5. John Mongillo, *Nanotechnology* 101 (Pentogon Press, 2007).
6. Smith, A.L., *Applied Infrared Spectroscopy* (Wiley: New York, 1979).
7. Griffiths, P.R. and De, Haseth J.A., *Fourier Transform Infrared Spectroscopy* (New York: John Wildey & Sons, 1986).
8. Atalla, R.H. and Agarwal, U.P., *Science* 227, 636-38 (1985).
9. Atalla, R.H. and Agarwal, U.P., *J. Raman Spectroscopy,* 17, 229-31 (1986).
10. Rao, C.N.R., *Ultraviolet and Visible Spectroscopy*, 2nd Ed., (Butterworths: London, 1967).
11. Julian, Chen C., *Introduction to Scanning Tunnelling Microscopy* (New York: Oxford University Press, 2008).

12. Stroscio, J.A. and Kaiser, W.A., *Scanning Tunnelling Microscopy* (Academic Press, 1993).
13. Williams, D. and Carter, C.B., *Transmission Electron Microscopy* (Plenum Press, 1996).
14. Fultz, B. and Howe, J., *Transmission Electron Microscopy and Diffractometry of Materials* (Springer, 2007).
15. Spence, John C.H., *Experimental High-resolution Electron Microscopy* (New York: Oxford University Press, 1988).
16. Joseph, G., Dale, E.N., David, C.J., Patrick, E., Charles, E.L. and Eric, L., *Scanning Electron Microscopy and X-ray Microanalysis*, 3rd Ed. (Springer, 2003).
17. Kourosh, Kalander-zadeh and Benjamin, Fry, *Nanotechnology Enabled Sensors* (New York: Springer, 2008).
18. Pradeep, T., *Nano: The Essentials* (New Delhi: Tata McGraw Hill, 2007).

6

Carbon Nanostructures

6.1 Introduction

Carbon has many useful properties which can be explored by suitably arranging the carbon atoms. It is one of the most versatile elements present in the earth. It is also a unique element among electronic materials as regards to its structure or property relations. It can be used as a metal (quasi-2D graphite), a semiconductor, a superconductor, a polymer, a quantum dot fullerene, or a quantum wire CNT. Carbon based nanomaterials are defined as materials in which the "nanocomponent" is pure carbon. Table 6.1 lists the carbon-based nanomaterials which are under investigation by the scientific community.

Carbon has been used for the reduction of metal oxides since B.C. 4000. In 1779, graphite, the allotropic form of carbon, was discovered. Ten years later another allotropic form of carbon, diamond was discovered. It was then found that both of these forms belong to a family of chemical elements. Kroto, Smalley and Curl discovered buckminsterfullerenes or fullerenes in 1985.[1-4] They were awarded Nobel Prize in Chemistry (1996) for this work. Sumio Iijima luckily discovered carbon nanotubes (CNTs) in 1991, while searching for new carbon structures in the deposits formed on graphite cathode surfaces during the electric arc-discharge.

6.2 Allotropes of Carbon

Allotropes of carbon are very stable due to the unique chemical bonding properties of carbon. Both the graphite and diamond are its natural allotropes, while fullerenes and their derivatives are manufactured in the laboratory. Allotropes of carbon are shown in figure 6.1. Physical and chemical properties of the allotropes of carbon are presented in Table 6.2. Carbon's allotropes have a wide range of amazing properties like high tensile strength and high melting points.

Fullerenes are molecules of varying sizes composed entirely of carbon in the shape of a cylindrical tube, hollow sphere, or ellipsoid. Spherical fullerenes are popularly known as buckyballs and have the formula C_{60}. Fullerenes have found tremendous applications in nanotechnology.

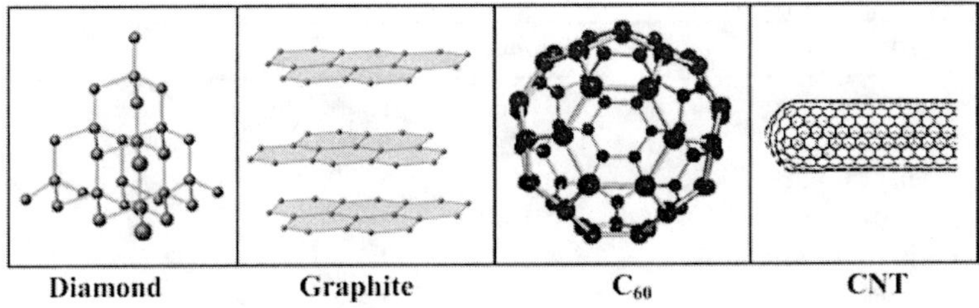

Fig. 6.1: Allotropes of carbon.

Table 6.1: Carbon-based nanomaterials[5-7]

Single Particles	Nanostructures Nanotubes	Films/Coatings/ Nanostructured surfaces	Nanostructured bulk material
Carbon Black*	SWNT	Carbon films	Nanostructured carbon
Fullerenes	MWNT	Diamond like carbon (DLC)	Nanoporous carbon
Graphite	Nanohorn	Covalent carbides like SiC	Carbon foams
Nanocluster	Nanowires	Metallic carbides like TiC	Carbon aerogels
	Nanorods	Nano Carbon Nitrides	Carbon nanocrystals

* Carbon black is the most widely used carbon nanomaterials now-a-days, it has applications in car tyres, antistatic textiles and is used for colour effects.

Carbon nanotube has a cylindrical shape and hence is also called buckytube. One end of the CNT is closed with a hemisphere of the buckyball structure. The length of the CNT is several centimetres whereas its diameter is only a few nanometres. CNTs are of two types: single-walled nanotubes (SWNTs) and multi-walled nanotubes (MWNTs). Their remarkable properties like high strength, excellent electrical and thermal properties offer potential applications in nanoelectronics, optics, sensors, energy storage and so on.

Table 6.2: Comparison of physical and chemical properties of the allotropes

Properties	Diamond	Graphite	C_{60}	CNTs
Colour	Colourless	Steel black to grey	Black solid/ Magenta in solution	Black
Density (g/cm^3)	3.515	1.9 – 2.3	1.69	1.33 – 1.4
Specific Gravity	3.52	2.2	1.7 – 1.9	2
Hardness (Moh's Scale)	10	1 – 2	1 – 2	1 – 2
Melting Point (°C)	3550	3652 – 3697	>800 (sublimes)	3652 – 3697
Boiling Point (°C)	4827	4200	—	—
Electrical Conductivity	Insulator	Conductor	Semiconductor	Conductor - Semiconductor
Hybridization	Sp^3-tetrahedral	Sp^2-trigional planar	Sp^2-trigional planar	Sp^2-trigional planar
Crystal Structure/ Shape	Cubic	Tabular	Truncated icosahedron	Cylindrical

6.3 Fullerene C_{60}

The fullerene C_{60} was discovered accidently. Smalley and Curl developed a technique to analyze atom clusters produced by laser vapourization with time of flight mass spectrometry (figure 6.2), which caught Kroto's attention because he was interested in the study of long-chain polyynes formed by red giant stars. When they used a graphite target, they could produce and analyze the long-chain polyynes. In September, 1985, they experimented with the carbon plasma, confirming the formation of polyynes. They observed two mysterious peaks at mass 720 and 840, corresponding to 60 and 70 carbon atoms, respectively as shown in figure 6.3. Further reactivity experiments determined a most likely spherical structure, leading to the conclusion that C_{60} is made of 12 pentagons and 20 hexagons arranged to form a truncated icosahedrons.[1-7] These investigators found Cn clusters, where n is even number and n>20. Fullerenes with larger number of carbon atoms such as C_{70}, C_{76}, C_{80}, C_{240} and even up to C_{540}, were also discovered.

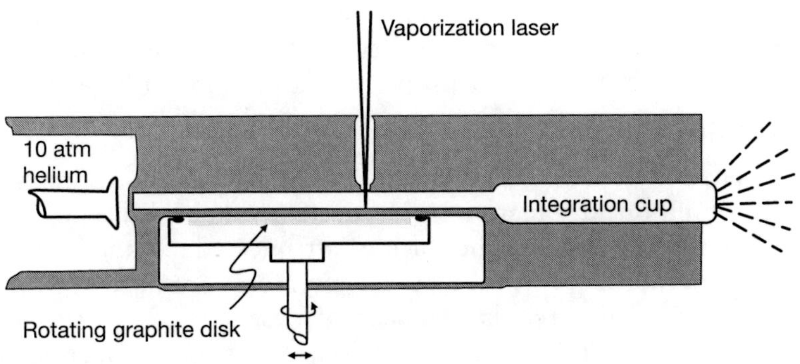

Fig. 6.2: Schematic diagram of the pulsed supersonic nozzle used to generate carbon cluster beams.

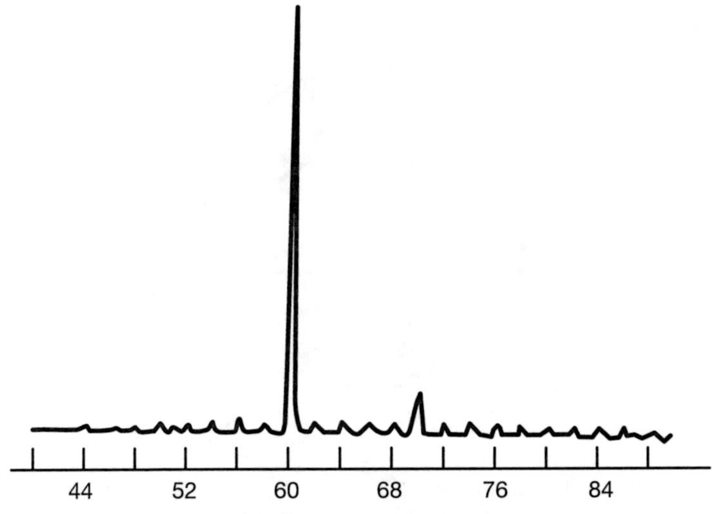

Fig. 6.3: Mass spectra of carbon clusters.

During the synthesis of fullerenes by any fabrication techniques, the most abundant among the fullerene species is C_{60}. Its stability is high and possess greatest symmetry. Almost all C_{60} molecules are identical. However in some cases, ^{13}C isotope (1.1% natural abundance) can substitute randomly for ^{12}C (98.9% abundance) in the caged molecule. The C_{60} molecule has monodisperse nanostructure with high (icosahedral I_h) symmetry. The fullerene C_{60} is the most interesting fullerene nanostructure due to its simplicity and relatively high abundance. The constituent fullerene molecules in a fullerene solid (fullerite) determine the structure or property relationships of fullerites.

6.3.1 Structure of C_{60}

The fullerene C_{60}, the molecule with the highest symmetry (I_h), is built of carbon atoms with mostly sp^2-hybridization. The atoms are assembled in the form of a truncated icosahedron as shown in figure 6.4. All the 60 carbon atoms are positioned at the vertices of a regular truncated icosahedron (0.710 nm diameter). It is important that all carbon sites are identical, consistent with a sharp line in the NMR spectrum. The average distance between nearest carbon-carbon (C-C) atoms (a_{c-c}) is extremely small about 0.144 nm, almost identical with graphite a_{c-c} distance (0.142 nm).

Fig. 6.4: Structure of the fullerene C_{60} molecule.[3]

In fullerene C_{60} molecule, carbon atoms at each corner is triagonally bonded to three other nearby carbon atoms, as in graphite. Out of the 32 faces of the regular truncated icosahedron,

20 faces are hexagons and the remaining 12 faces are pentagons. Thus, C_{60} molecule may be considered as a "rolled-up" graphene sheet (a single layer of crystalline graphite), which forms a closed shell molecular nanostructure, obeying Euler's theorem. Euler's theorem states that a closed surface consisting of hexagons and pentagons has exactly 12 pentagons, and an arbitrary number of hexagons. The introduction of pentagons gives rise to curvature in forming a closed structure of the molecule as shown in figure 6.5(a). To minimize local curvature, the pentagons become separated from each other in the self-assembly process, giving rise to the isolated pentagon rule (IPR), an important rule for stabilizing fullerene clusters. IPR states that a stable and non-reactive fullerene is only formed when the pentagons at its surface are separated. Thus, C_{60} is the smallest stable fullerene formed, and it has only one isomer. Smaller, non-IPR structures like C_{20} and C_{36} have been produced successfully. However, they can only be studied in the gas phase or in an oligomerized/ polymerized form in the solid due to their high reactivity. The smallest possible fullerene structure that obeys Euler's theorem is C_{20} which would form a regular dodecahedron with 12 pentagonal faces. But this non IPR structure is unfavourable because of its high local curvature and strain. The addition of a single hexagon adds two C atoms, thus all fullerenes C_{nc} contain an even number of carbon atoms (n_c), in agreement with the observed mass spectra for fullerenes.

The diameter (di) of a fullerene can be estimated from the relation for an icosahedral fullerene,

$$di = a_{c-c}\,[(15n_c)^{1/2}]/2\pi \qquad ...(6.1)$$

where $a_{c-c} = 0.144$ nm is the average nearest-neighbour carbon-carbon distance. If $n_c < 10^3$, the diameter of the fullerene C_{nc} nanostructure di ≤ 3 nm. Although all C atoms in C_{60} are equivalent, the bonds between different atoms are not equivalent. The valency of a carbon atom is four. These valence electrons of every carbon atom is engaged in covalent bonds, so that the two bonds on the pentagon perimeter are electron-poor single bonds ($a_{c-c} = 0.146$ nm), and the bond between two hexagons is an electron-rich double bond ($a_{c=c} = 0.140$ nm). Since each carbon atom has completed octet, the C_{60} molecule is likely a van der

Waals bonded crystal which is nonconducting (an insulator or a semiconductor). Upon condensation into a solid, the C_{60} molecules form a close-packed structure with face centred cubic (f.c.c.) symmetry and a lattice constant (a_0 = 1.4198 nm) under ambient conditions.

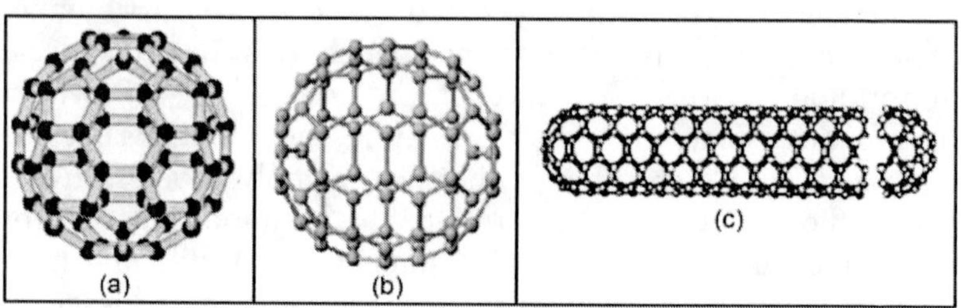

Fig. 6.5: Examples of closed shell fullerene configurations: (a) C_{60}, (b) C_{70}, and (c) an armchair carbon nanotube.

Source: www.upload.wikimedia.org; www.planetseed.com

6.3.2 Structure of Higher Fullerenes

Number of larger molecular weight fullerenes (C_{nc}, n_c > 60) are also developed during the synthesis of C_{60}. The C_{70} is the most abundant among them. It is possible to isolate significant amounts of higher mass fullerenes like C_{70}, C_{76}, C_{78}, C_{80}, C_{82}, and so on. These higher mass fullerenes also possess a close-packed crystal structure. Because of the lower molecular symmetries of higher fullerenes, their symmetry, molecular orientation and temperature dependence are more complicated than that of C_{60}. Some higher fullerenes show deviation from the quasi-spherical shape. The fullerene C_{70} molecule shows a rugby-ball shape, and is formed by adding a ring of 10 carbon atoms or a belt of 5 hexagons around the equatorial plane of the C_{60} molecule normal to one of the five-fold axes. Higher fullerene molecules (> C_{76}) exhibit a further degree of freedom because for each carbon atom there exists more than one structural isomer of the fullerene. The lattice constant 'a' of higher fullerene solids is proportional to $\sqrt{n_c}$, as depicted in figure 6.6. This behaviour arises because the average radius of the molecules is roughly proportional to $\sqrt{n_c}$ since the C–C bond lengths will not change appreciably with the size of the fullerene molecules.

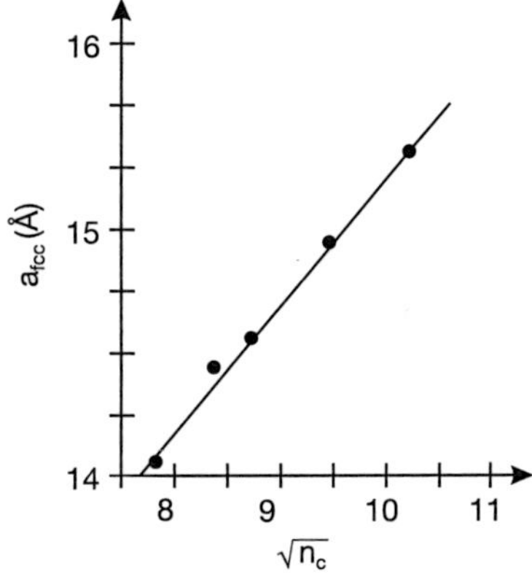

Fig. 6.6: Lattice constant (a_{fcc}) of some higher fullerenes (C_{60}, C_{70}, C_{76}, and C_{84}) as a function of $\sqrt{n_c}$.

Fullerenes often form isomers because a closed cage molecule C_{nc} can form different geometrical structures. Each structure corresponds to an isomer. When nc increases, the number of isomers obeying the isolated pentagon rule increases rapidly. For example, C_{78} has 5 isomers and C_{80} has 7 isomers. However, fullerenes C_{60} (I_h), C_{70} (D_{5h}) and C_{76} (D_2) have only a single isomer each, which obeys the isolated pentagon rule.

6.3.3 Doping of Fullerenes

There are mainly three doping routes in which the electronic properties of a fullerene can be modified in a controlled manner. The doping methods are (a) doping from outside the molecules (intercalation), resulting fullerene salts, (b) doping from inside the molecule (endohedral), resulting metallofullerenes, and (c) changing the molecular structure itself by doping on the buckyball called heterofullerenes, as illustrated schematically in figure 6.7. These doping methods are successful and can produce a number of fullerene intercalation compounds, endohedrally doped fullerenes (metallofullerenes) or heterofullerenes which substitute C atoms with heteroatoms like N or B. Besides, it is also possible to dope from outside through charge injection in a field effect device. All these techniques are successful in property engineering and thereby fullerene based materials offer a variety of properties like ferromagnetism, metallic conductivity, superconductivity, non-linear optical properties and so on.

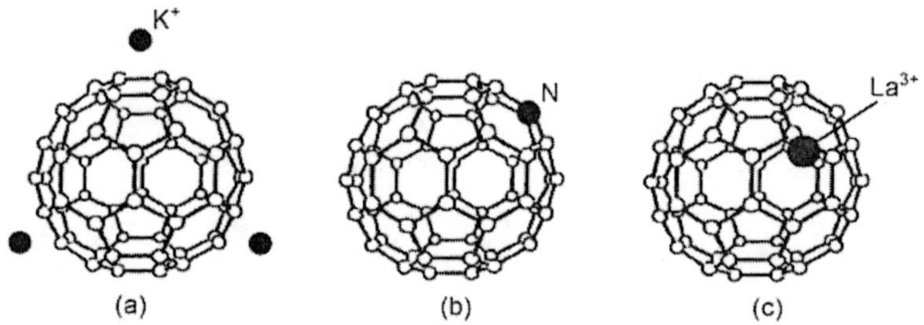

Fig. 6.7: Doping fullerenes: (a) from outside, (b) on-ball, and (c) from inside.[3]

If alkali (A) or alkaline earth (AE) metals dope with fullerene C_{60}, a number of intercalation compounds, called fullerides, are being constructed. These fullerides usually take up face centred cubic (f.c.c.), body centred cubic (b.c.c.) or body centred tetragonal (b.c.t.) crystal structures. The interstitial sites of the fullerene solids are occupied by the metal atoms. The metal atoms contribute their outer electron(s) to the C_{60} molecules. The arrangement of C_{60} and alkali ions in structures with intercalated compounds A_3C_{60}, A_6C_{60}, and A_4C_{60} is shown schematically in figure 6.8. The large and small spheres represent the C_{60} molecules and the alkali ions respectively.

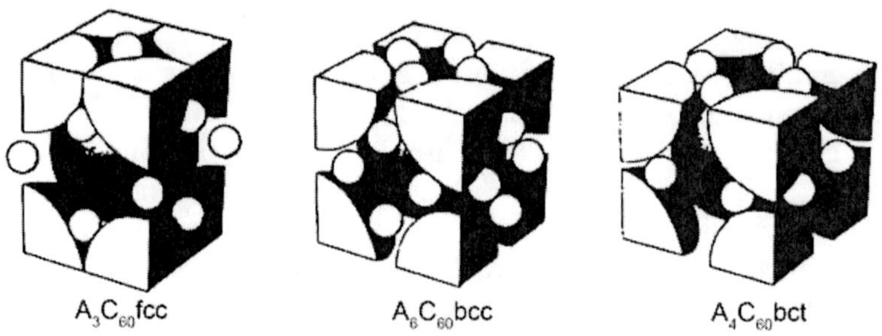

Fig. 6.8: Intercalated compounds: A_3C_{60}, A_6C_{60} and A_4C_{60} compounds.[8]

In 1991, heterofullerenes molecules, such as $C_{60-n}B_n$ and C_mN_n molecules were produced using molecular beam synthesis routes. The heterofullerene $C_{59}N$ was the first one made available in large scale for solid state spectroscopic studies. It forms dimers ($(C_{59}N)_2$) and crystallizes in a lattice with monoclinic symmetry. Besides, the intercalation of alkali metal with $C_{59}N$ is also possible, which combines two fullerene doping mechanisms simultaneously (combinational doping). $K_6C_{59}N$ was the first intercalated phase, where the dimer bonds are broken and the structure is b.c.c., similar to that of K_6C_{60}.

6.3.4 Metallofullerenes

As discussed in 6.3.3, metallofullerenes are endohedrally doped fullerenes in which metal dopant is doped from inside. Metallofullerene nanostructures having a metal dopant within the fullerene cage as shown in figure 6.9. Different metal species can be introduced into the interior hollow core of the endohedrally doped C_{60} molecule. Usually one, two, or three metal species can be put inside a single fullerene cage. The endohedral fullerene configuration is usually denoted as $M@C_{nc}$. For example, $La@C_{60}$ denotes one endohedral lanthanum in a C_{60} molecule and $Y_2@C_{82}$ denotes two Y atoms inside a C_{82} fullerene molecule.[9] It is important that the stability of an endohedral fullerene depends on many significant parameters such as the dopant species, the number of dopant atoms or ions, the number of carbon atoms, the amount of charge transfer between the fullerene cage and the dopant, and the shape of the fullerene isomer.

The location of the dopants inside the cage may not essentially be at the centre of the molecule as shown in figure 6.9. Metallofullerenes commonly exhibit dipole moment because the interaction between the dopant and the fullerene molecule may result charge transfer between them. Besides, these endofullerenes have large dipole moment (2-4 Debye) due to relatively large size of the fullerene shell. The high dipole moment may affect the solubility of specific metallofullerenes in solvents. This property would be exploited in isolating and purifying metallofullerenes. Moreover, other endohedral dopants (e.g. He and Ne) of fullerenes are also possible even with lower concentrations.

Possible structural models of $M@C_{60}$, $La@C_{82}$ and $Sc_3@C_{82}$ metallofullerenes with dopants at different locations are shown in figure 6.9.

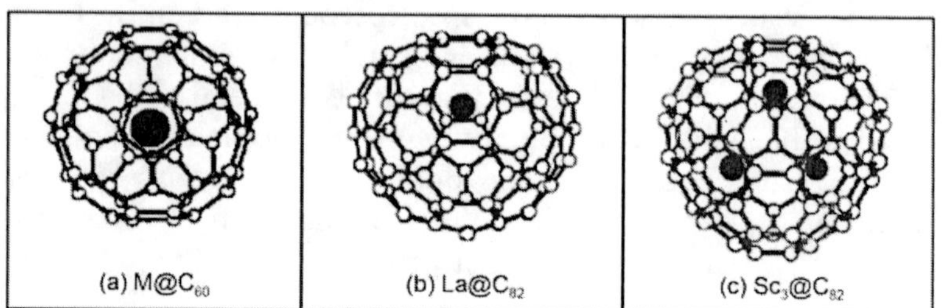

Fig. 6.9: Structural models of (a) $M@C_{60}$, (b) $La@C_{82}$, and (c) $Sc_3@C_{82}$ metallofullerenes (black balls represent the dopant).[9]

6.3.5 Metal Coated Fullerenes

Metal coated fullerene clusters can be fabricated by vapour synthesis method provided the metal exohedral dopants satisfy size limitations. Monolayer coatings of metal atoms are suitable for alkali metal coated fullerenes.[9] For example, in $Li_{12}C_{60}$, Li is supposed to be located in the centres of each pentagonal face of fullerene C_{60}. Alkaline earth atoms like Ca, Sr, and Ba are suitable to form multilayer metal structure over the C_{60} surface. For example, in $Ca_{32}C_{60}$ one Ca is placed over each of the 12 pentagonal and hexagonal faces. The multilayer metal structures of ordered Ca atoms over the fullerene C_{60} surface are shown schematically in figure 6.10. The Ca atoms positioned over the icosahedral vertices of C_{60} are darkened and N denotes the number of metal layers.

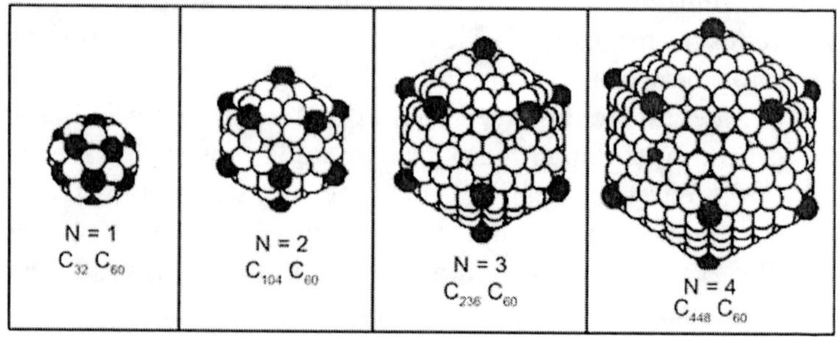

Fig. 6.10: Multilayer metal (Ca) covered fullerenes.[9]

6.3.6 Superconductivity in C_{60}

Superconductivity is a state of matter in which the electrical resistance of a sample becomes zero, and in which no magnetic field is permitted to penetrate the sample. This phenomenon is observed in several metals, ceramic materials and nanoscale materials. Kraus et al. (1995) have reported superconductivity in Ba intercalated C_{60}. Magnetization measurements of samples with a Ba_6C_{60} show superconductivity at a critical temperature (Tc) of 6.5 K. The superconductivity in bulk chemically intercalated fulleride salts with a critical temperature of 33 K and in hole doped C_{60} derivatives in field effect transistor (FET) configurations with a critical temperature of 117 K are reported in the literature.

6.3.7 Applications of Fullerenes

Fullerenes have found tremendous applications in nanotechnology. Fullerenes draw lot of attention from researchers and technologists worldwide because of their unique structures and properties. The scientific community has been trying to generate novel applications of these new carbon structures. Fullerenes and their derivatives find potential use in chemical sensors like quartz crystal microbalance (QCM) and surface acoustic wave sensors (SAW). Piezoelectric crystal is the main component of both QCM and SAW. This crystal is highly sensitive to mass variations on its surface, and hence it can be applied for trace quantitative analysis. Since fullerene molecule and its derivatives show affinity to organic molecules, they can be coated onto piezoelectric crystal. This property enables them to be used as chemical sensors for organic molecules like propanol, butanol, ethanol and methanol. The frequency of oscillation of piezoelectric crystal decreases as the organic molecules adsorbed onto the fullerene derivative film. This change of frequency depends on the concentration of the adsorbent and hence, the concentration of the adsorbent can be measured.

Fullerenes can be used in photovoltaic cells as organic photovoltaics (OPV). The efficiency of the fullerene/polymer blend bulk heterojunction polymer solar cell is high. In OPV, n-type fullerene and a p-type polymer, usually a polythiophene, are combined. They are blended and cast as the active layer to produce a bulk heterojunction. The semiconducting properties of C_{60}, C_{70}, or C_{84} can be used for the construction of Organic Field Effect Transistors (OFETs). OFETs made with C_{84} exhibit greater mobility and stability than that made with C_{60} or C_{70}.

Fullerenes have very useful applications in health care. For example, it helps to prevent oxidative cell damage. Fullerenes based drugs can be used in controlling neurological damages caused by diseases. For example, Alzheimer's disease is the result of radical damage. Pharmaceutical companies are also trying to explore the use of fullerene drugs in many areas like photodynamic therapy, anti-viral agents and so on.

Fullerenes and fullerenic black are suitable to mix with polymer structures to create co-polymers and nanocomposites since they are chemically reactive. The new materials show enhanced properties. Other applications of fullerenes include catalysts, effective water purification and biohazard protection.

6.4 Carbon Nanotube Structures

CNTs are concentrically rolled graphene sheets with a large number of potential helicities and chiralities.[10-15] Iijima initially observed only MWNTs with around 2-20 layers. Later he investigated single walled carbon nanotubes (SWNTs) and explained their structure.

SWNTs have single sp^2 hybridized carbon sheets rolled into seamless tubes. The way in which the sheet is rolled determines the fundamental properties of the tube. The diameters of SWNT ranges from 0.7 to 3 nm and its length to diameter ratio is about 1000. Thus, CNTs are generally considered to have one-dimensional structures. A SWNT consists of a sidewall and end caps. The physical and chemical properties of these regions are different. The structure of the end cap is similar to fullerene C_{60}. The structural details of C_{60} are presented in section 6.3.1.

The side wall tube of SWNT is generated when a graphene sheet of suitable dimension is wrapped in a specific direction. One of the two atoms chosen in the graphene sheet can be treated as the origin. The sheet is rolled up till the two atoms coincide. The vector pointing from the origin to the second atom is called the chiral vector. The length of the chiral vector is equal to the circumference of the nanotube (figure 6.11). The direction of the chiral vector is perpendicular to the tube axis. SWNTs with different chiral vectors show different physical properties, such as optical, mechanical and electrical properties.

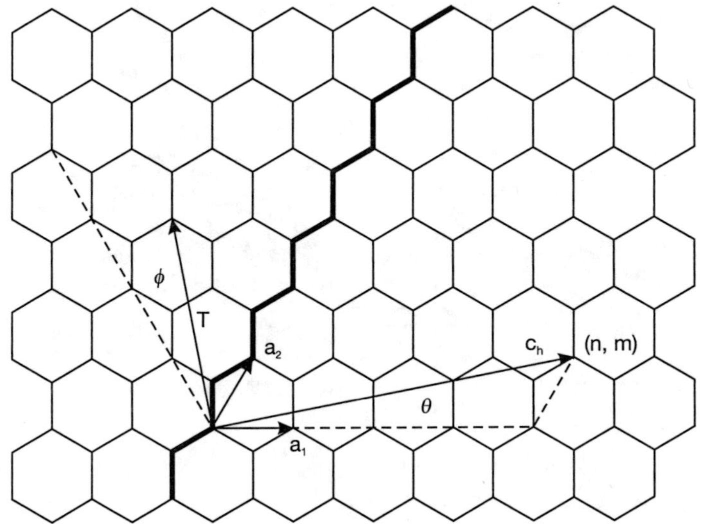

Fig. 6.11: Structure of two-dimensional graphene sheet with chiral vector C_h, unit vectors a_1 and a_2 and the chiral angle θ.

A nanotube is produced when a graphite sheet is rolled up about the axis T, as shown in figure 6.11. To describe fundamental characteristics of the nanotube, two vectors, C_h and T, can be introduced. C_h is the vector that defines the circumference on the surface of the tube connecting two equivalent carbon atoms. The chiral vector $C_h = na_1 + ma_2$, where a_1 and a_2 are the two basis vectors of graphite and n and m are integers.

The chiral angle $\theta = \tan^{-1} [\sqrt{3}n/(2m + n)]$ ...(6.2)

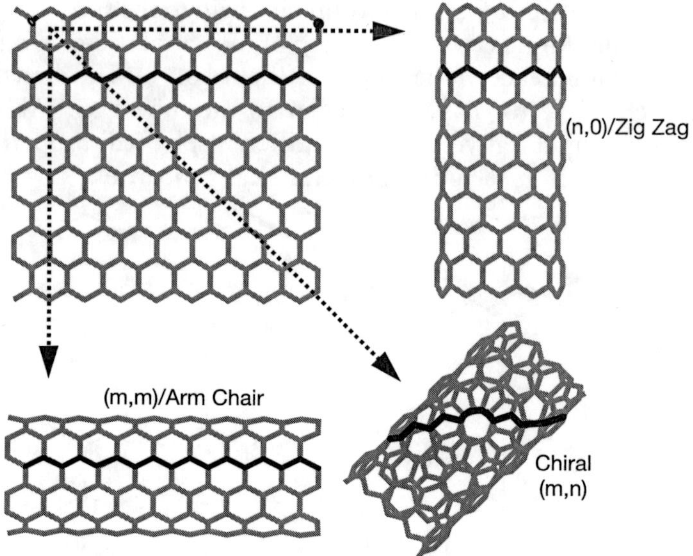

Fig: 6.12: Schematic illustration of rolling graphite sheet to create a CNT.[16]

The chiral angle is used to separate carbon nanotubes into three classes differentiated by their electronic properties: armchair ($n = m$, $\theta = 30°$), zig zag ($m = 0$, $n > 0$, $\theta = 0°$), and chiral ($0 < |m| < n$, $0 < \theta < 30°$), as shown in figure 6.12. Armchair carbon nanotubes are metallic (a degenerate semimetal with zero band gap). Zig zag and chiral nanotubes can be semimetals with a finite band gap if $(n - m)/3 = l$($l = 0, 1, 2, \ldots$ and $m \neq n$) or semiconductors in all other cases. Figure 6.12 illustrates how different CNTs are made from graphite sheet. The unique electronic behaviour of each nanotube depends on its band gap. The band gap of the semimetallic and semiconductor nanotubes depends inversely as their diameter. Moreover, the electronic properties of CNTs depend on diameter and chirality.[17-24]

The diameter of the nanotube can be expressed as

$$d_t = \sqrt{3}[a_{c\text{-}c}(m^2 + mn + n^2)^{1/2}/\pi] = C_h/\pi \qquad ...(6.3)$$

where C_h is the magnitude of $\mathbf{C}_h$, and $a_{c-c} = 1.42$ Å, the C–C bond length. Combining different diameters and chiralities may result several hundred individual nanotubes, each with its own distinct mechanical, electrical, piezoelectric and optical properties. The three types of SWNTs with armchair, zig zag and chiral structures are shown in figure 6.13.

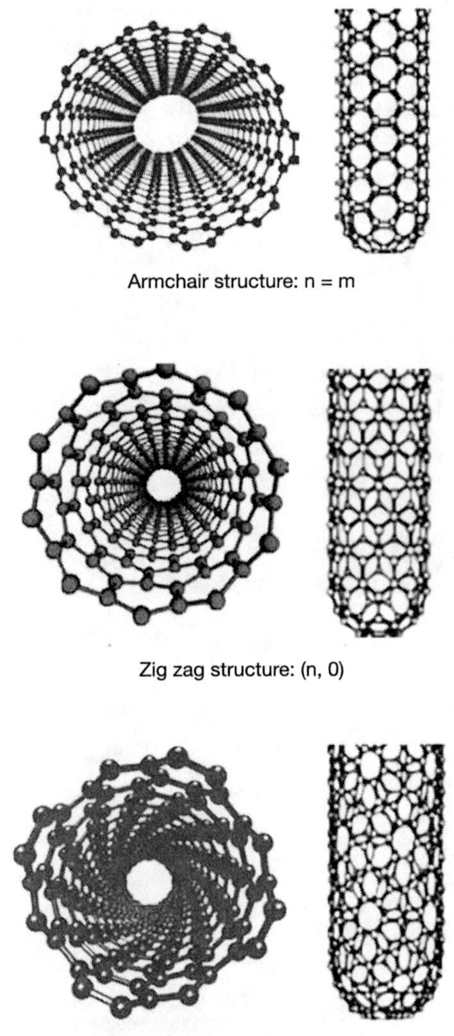

Fig. 6.13: Three types of SWNTs identified by the integers (n, m).

6.4.1 Physical Properties of Carbon Nanotubes

6.4.1.1 Electrical Properties

CNTs have remarkable electrical properties. The suitable combination of the structural parameters n and m can create metallic CNTs of appropriate conductivity. The values of n and m determine degree of twist of the nanotube. As discussed in section 6.4, the electrical conductivity of nanotube depends on both diameter and degree of twist (chirality). Hence, CNTs can be made metallic or semi-conducting, which depends on chirality. Even a small variation of chirality can change a metallic CNT into a semiconductor one.[25-26] Carbon nanotubes with different twisting angles are shown in figure 6.14(a) and the effect of twisting on the properties of a metallic nanotube is shown in figure 6.14(b).

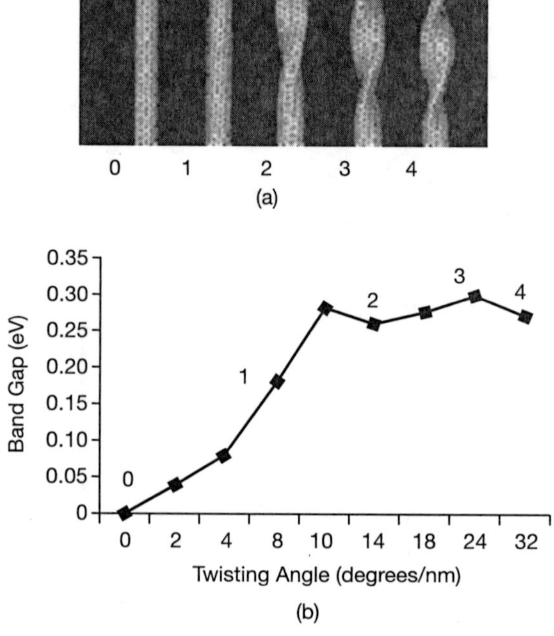

Fig. 6.14: (a) CNTs with different twisting angles and (b) Variation of band gap energy with twisting angle.[25]

A plot of energy gap of the semiconducting nanotubes versus diameter is shown in figure 6.15. The graph shows that as the diameter of the tube increases, the band gap decreases. The expression for band gap

$$E_{gap} = (2y_0 a_{cc})/d \quad ...(6.4)$$

where y_0 is the C–C tight bonding overlap energy (2.7 ± 0.1 eV), a_{cc} is the nearest C–C distance (= 0.142 nm) and d the diameter of the nanotube. The fundamental energy gap ranges from 0.4 eV - 0.7 eV. Various investigations have established that the band gap of a semiconductor nanotube changes with small variations of diameter and bond angle.

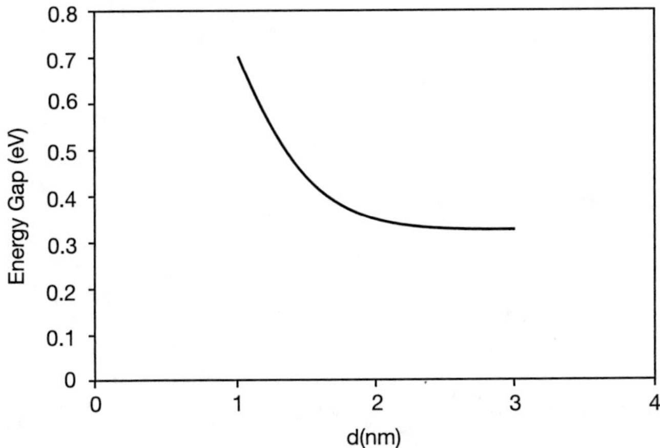

Fig. 6.15: Variation of energy gap with diameter.

The conductance study of nanotubes found that the nanotube behaved as a ballistic conductor with quantum behaviour. The value of conductance, G_0 was found to be 1/12.9 $k\Omega^{-1}$, where $G_0 = 2e^2/h$. In the metallic state the conductivity of the nanotube is very high. The electrical properties of ropes of SWNTs can be investigated by connecting electrodes at different parts of the nanotubes. The resistivity measured for metallic SWNT is approximately 10^{-6} Ω–m at room temperature. Thus, SWNT ropes are the best conductive carbon fibres. Besides, the current density reported for CNT is greater than 10^{11} A/m^2. A single carbon nanotube can carry up to 25 μA of current. This corresponds to an extraordinarily high current density of 10^{13} A/m^2. A typical V – I characteristic for a metallic carbon nanotube is shown in figure 6.16. There is a linear increase of current at low bias voltage (the ballistic regime), but then rolls over and saturates at high bias.

MWNTs show superconductivity with interconnected inner shells, at a critical temperature (T_c) of 12 K. However, the critical temperature is lower for ropes of SWNTs and MWNTs without interconnected shells. Ultra small SWNTs exhibit superconductivity at a transition temperature of about 20 K.

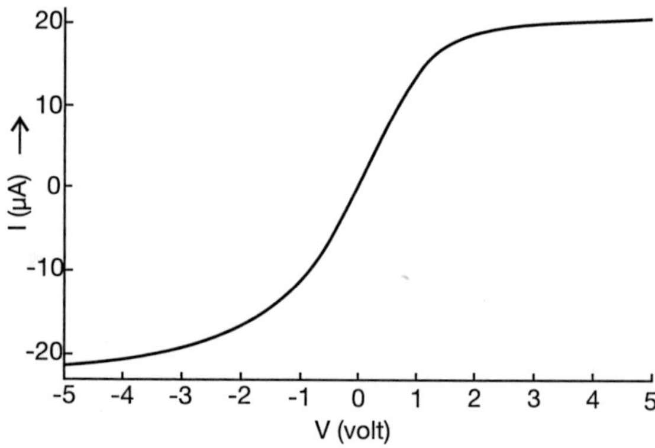

Fig. 6.16: V–I Characteristics of a metallic carbon nanotube.

It is interesting that individual SWNTs contain defects and these defects help the SWNTs to work as transistors. Besides these, connecting CNTs together may help to develop new transistor like devices. For example, if straight metallic section of a SWNT is joined to a chiral semiconducting section, a diode will be formed. When SWNTs are used as interconnects on semiconductor devices, they can transmit electrical signals at an amazing speed of about 10 GHz. It is worth to note that semiconductor SWNTs can be used in field effect transistors (FETs) instead of silicon. Furthermore, the energy dissipation in nanotubes will be very small because of their low resistance.

6.4.1.2 Mechanical Properties

Carbon nanotubes have very good mechanical properties because of the strength of the sp^2 carbon-carbon bonds. CNTs are considered to be the strongest and stiffest materials in terms of their tensile strength and elastic modulus. In addition, they are extremely resistant to damage and can withstand physical forces. When stress is applied on the tip of a nanotube, it will bend without any damage to the tip. The tip regains its initial state just after the deforming force is removed. This property of CNTs makes them suitable for probe tips in high resolution scanning probe microscopy.

The typical value of Young's modulus of nanotubes is 1-2 TPa, about 5 times higher than steel. The tensile strength of nanotubes is around 150 GPa, about 50 times higher than steel. These two properties along with the lightness of CNTs, give them great potential in applications like aerospace. The fabrication of a super hard material with a hardness of 62-152 GPa by compressing SWNTs over 24 GPa at 300 K has been reported in the literature. The bulk modulus (K) of the super hard material is 462-546 GPa, whereas the K value of diamond is 420 GPa only. These exceptional mechanical properties are ideal for reinforced composites, nanoelectromechanical systems (NEMS) and for probes in scanning force microscopes.

Table 6.3: Physical properties of carbon nanotubes

Material	Density (g/cm^3)	Young's Modulus (TPa)	Tensile strength (GPa)	Elongation at break (%)	Thermal Conductivity W/m/K
SWNT	-	~1 (from 1 to 5)	13–53	16	~2000
Armchair SWNT	1.33	0.94	126.2	23.1	-
Zig zag SWNT	1.34	0.94	94.5	15.6–17.5	-
Chiral SWNT	1.40	0.92	-	-	-
MWNT	-	0.8–0.9	11–150	-	-

6.4.1.3 Magnetic Properties

Carbon nanotubes, either metallic or semiconducting, possess novel magnetic properties. It is significant that by applying suitable magnetic field along the axis of a carbon nanotube, its metallicity can be controlled. A CNT can be made either semiconducting or metallic, depending on the strength of the applied field. Its band gap will be an oscillatory function of magnetic field with a period $\Phi_0 = h/e$ (the magnetic flux quantum). Thus, metallic tubes can be made semiconducting by applying a (even infinitesimally small) longitudinal magnetic field. But, the semiconducting nanotubes can become metallic in ultrahigh magnetic fields.

Systematic investigation on the magnetic susceptibilities of multiwalled carbon nanotubes as a function of temperature or magnetic field has found that the susceptibility of CNT was only around half that of graphite. It is also found that per carbon basis the susceptibility of CNT bundles is even larger than that of graphite. Studies of the magnetic properties of aligned multiwalled carbon nanotubes at various temperatures and magnetic field orientations have found that the nanotubes are diamagnetic and anisotropic. Carbon nanotubes show variation in its resistance with a DC magnetic field at low temperature. This effect is called magnetoresistive effect. The transition of the samples from positive to negative magnetoresistance is also reported.

6.4.1.4 Aspect Ratio

CNTs exhibit high aspect ratio (about 1000:1), and hence can be used as conductive additive materials for all types of plastics. This large value of aspect ratio helps to provide electrical conductivity in plastics at small concentrations. Furthermore, CNTs have the potential to become an outstanding additive to offer electrical conductivity in plastics.

6.4.2 Applications of Carbon Nanotubes

The novel properties of CNTs make them ideal candidate for many potential applications such as energy storage, actuators or artificial muscles, AFM probe tips, batteries, biosensors, drug delivery, data storage, hydrogen storage, microelectromechanical (MEMS) devices, nanolithography, super capacitors, thermal protection, waste recycling, composite materials, nanoporous filters and so on.[27-33] A few potential applications of carbon nanotubes are presented in this section.

6.4.2.1 Molecular Electronics and Integrated Circuits

In any electronic circuit, particularly at nanoscale dimensions, the interconnections between various active and passive components are very difficult. The geometry, electrical conductivity and other unique characteristics of CNTs make them suitable for the interconnections in molecular electronics, as shown in figure 6.17. Moreover, they can be used as switches themselves. The field effect transistor (FET) can be constructed with a semiconducting SWNT. When a suitable voltage is applied to the gate of the FET, the conducting nanotube transforms to a non-conducting state. Such CNT transistors can be coupled together to function as a logic gate, the basic component of computers. In 1998, the first carbon nanotube field effect transistors (CN-FETs) were fabricated at Delft and at IBM.

CNTs have very high thermal conductivity and hence, they can be used to develop CNT based heat sinks to eliminate heat from ICs and microprocessors. Since CNTs have novel conducting and semiconducting characteristics, they are robust for electronic devices rather than silicon based electronic devices. Hence, individual CNT based electronic devices may replace silicon devices.

Fig. 6.17: MWNT Interconnects.[31]

6.4.2.2 Field Emission and Flat Panel Displays

When a small electric field is applied along the nanotube axis, the ends of CNT emit electrons with a large emission rate like water being pushed through a high powered hose. This effect is called field emission. The emission of electrons increases as the strength of the electric field enhances. The emission also depends on work function of the material. Carbon nanotubes are the best field emitters of any known materials because of the tip sharpness and large electrical conductivity. Even with a small p.d. applied across the tip and an electrode that is close to the tip, a very high electric field (nearly millions of volts/cm) set up near the tip. This high electric field is due to the tip sharpness, and the electric field generated is inversely proportional to the radius of curvature of the tip. These large electric fields pull an enormous number of electrons out of the tip. For example, CNT films emit nearly 4 A/cm^2. Furthermore, the emission current is extremely stable. These significant properties of CNTs are employed in the construction of field emission flat panel displays. For each pixel in the display of a flat panel, separate electron guns are used instead of a single electron gun in the conventional CRT display. The unique features of CNTs such as, high current density, low operating voltage, steady and long life make them ideal field emitters for flat panel displays. All these features make CNTs appropriate for many other applications also, such as AFM tips, microwave amplifiers, lightning arrestors, electron microscope cathodes, high resolution X-ray sources and cold-cathode lighting sources.

Researchers are looking ahead to explore suitable technology to replace conventional electron gun with CNT electron gun. Figure 6.18 shows the schematic diagram of a CNT based flat panel display. Such flat panel displays consume less power; exhibits speed and excellent brightness; offer wide viewing angle and a large operating temperature range.

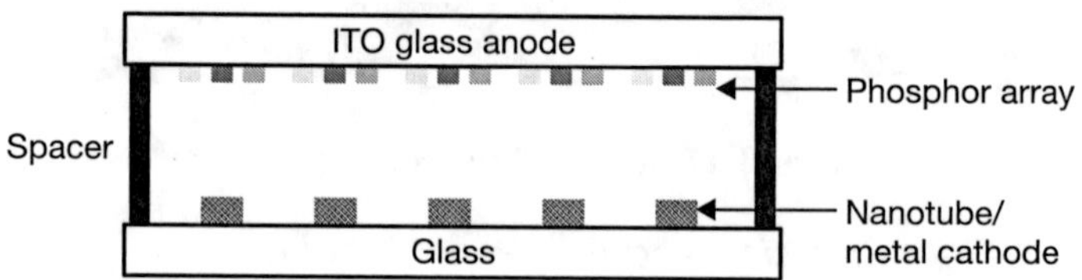

Fig. 6.18: Schematic illustration of a CNT based flat panel display. (ITO: indium tin oxide)

6.4.2.3 Medical Applications

Nanotubes have a wide range of significant applications in medical field. They find profound impact on biomedical and pharmaceutical fields. Numerous investigations have established that CNTs could be used as a drug delivery vehicle to carry drugs to infected cells. In the near infrared (NIR) range, single walled carbon nanotubes show strong light absorbance. When this light illuminates SWNT, localized heat will be produced due to the absorbance of NIR light. The heat thus produced will help to stimulate the release of drugs or genes from the nanotube surface. Multiwalled nanotubes can easily pass through the cell membrane even without cell damage, as they act like nanoneedles. CNT based nanobiosensors can be used for diagnostics and drug discovery. Besides, CNTs can act as a powerful tool to explore basic information about targets for therapies, since they can detect small amounts of molecules through electrical, optical, or mechanical means.

Actuators or Artificial Muscles

CNT actuators convert electrical energy to mechanical energy, causing the nanotube to move. Usually, a CNT actuator consists of two pieces of paper made from CNTs and are placed on either ends of a piece of tape that is attached to an electrode through which electric current is passed. When current is applied, the electrons will be pumped into a piece of CNT paper. Hence, nanotubes on that side of the paper expand and cause the tape to curl in one direction. This movement is similar to the way an artificial muscle works. Nanotube actuators have applications in robotics and prosthetics.

REFERENCES

1. Dresselhaus, M.S. and Dresselhaus, G., "Nanotechnology in Carbon Materials", *Nanostructured Materials* 9, 33-42 (1997).
2. Dresselhaus, M.S., Dresselhaus, G., and Saito, R., *Carbon* 33, 883-91 (1995).
3. Martin, Knupfer, "Electronic Properties of Carbon Nanostructures", *Surface Science Reports*, 42, 1-74 (2001).

4. Poole, C.P. and Owens, F.J., *Introduction to Nanotechnology* (John Wiley & Sons, 2006).
5. Dresselhaus, M.S., Dresselhaus, G. and Eklund, P.C., *Science of Fullerenes and Carbon Nanotubes* (NY: Academic Press, 1996).
6. Bharath Bhushan, *Handbook of Nanotechnology* (NY: Springer, 2004).
7. John, Mongillo, *Nanotechnology* 101 (Pentagon Press, 2007).
8. Fleming, R.M., Rosseinsky, M.J., Ramirez, A.P., Murphy, D.W., Tully, J.C., Haddon, R.C., Seigrist, T., Tycko, R., Glarum, S.H., Marsh, P., Dabbagh, G., Zahurk, S.M., Makhija, A.V. and Hampton, C., *Nature* 352, 701 (1991).
9. Dresselhaus, M.S., Dresselhaus, G., and Eklund, P.C., *Science of Fullerenes and Carbon Nanotubes* (NY: Academic Press, 1996).
10. Michael, J. (ed.), Carbon nanotubes: properties and applications (Taylor & Francis, 2006).
11. Popov, V.N. and Lambin, P. (ed.), *Carbon Nanotubes* (The Netherlands: Springer, 2006).
12. Kroto, H.W., Heath, J.R., Brien, S.C.O., Curl, R.F. and Smalley, R.E., *Nature*, 318, 162 (1985).
13. Iijima, Sumio, *Nature* (London, United Kingdom), 354, (6348), 1991.
14. Ajayan, P.M. and Ebbesen, T.W., *Rep. Prog. Phys.*, 60, 1025-65 (2003).
15. Takikawa, Hirofumi, Tao, Yoshitaka, Hibi, Yoshihiko, Miyano, Ryuichi, Sakakibara, Tateki, Ando, Yoshinori, Ito, Shigeo, Hirahara, Kaori, and Iijima, Sumio, AIP Conference Proceedings, 590, (Nanonetwork Materials), 2001.
16. Meyyappan, M., *An Overview of Recent Developments in Nanotechnology*, NASA Ames Research Center, (www.ipt.arc.nasa.gov).
17. Takikawa, H., Ikeda, M., Hirahara, K., Hibi, Y., Tao, Y., Ruiz, P.A., Sakakibara, T., Itoh, S., and Iijima, S., *Physica B: Condensed Matter* (Amsterdam, Netherlands), 323, (1-4), 2002.
18. Reynhout, X.E.E., Reijenga, J.C. and Daenen, M., *The Wondrous World of Carbon Nanotubes* (Eindhoven University of Technology, 2003).
19. Paul, L. McEuen, Marc, Bockrath, David, H. Cobden, and Jia, G. Lu, *Microelectronic Engineering* 47 (1999) 417-20.
20. Hiroshi, Ajiki and Tsuneya, AndoHiroshi, *Solid State Communications*, Vol. 102, No. 2-3, pp. 135-42, 1997.
21. Ajayan, P.M., *Prog. Crystal Growth and Charact.* 34, 37-51 (1997).

22. Bonard, J.M., Kind, H., Thomas, S. and Nilsson, L.O., *Solid State Electronics* 45, 893-914 (2001).

23. Wang, Z.L., Gao, R.P., Poncharal, P., de Heer, W.A., Dai, Z.R., Pan, Z.W., *Materials Science and Engineering* C 16, 3-10 (2001).

24. Vivien, L., Lanc, P., Riehl, D., Hach, F., and Anglaret, E., *Carbon* 40, 1789-97 (2002).

25. Balakrishna, K.M. and Varghese, Thomas, *Sci. & Soc.* 3(2), 11 (2005).

26. Kong, J. and Dresselhaus, M.S., *LabPlus International* 18, 14 (2004).

27. Chae, H.G., Minus, M.L., Rasheed, A. and Kumar, S., *Polymer* 48, 3781 (2007).

28. Choi, W.B., Chung, D.S., Kang, J.H., Kim, H.Y., Jin, Y.W., Han, I.T., Lee, Y.H., Jung, J.E., Lee, N.S., Park, G.S., and Kim, J.M. *Applied Physics Letters* 75, 3129-31 (1999).

29. Dai, H., Hafner, J.H., Rinzler, A.G., Colbert, D.T. and Smalley, R.E., *Nature* 384, 147-50 (1996).

30. de Heer, W.A., Châtelain, A. and Ugarte, D., *Science* 270, 1179-80 (1995).

31. Iijima, S., *Nature* 354, 56 (1991).

32. Kroto, H.W., Heath, J.R., Obrien, S.C., Curl, R.F. and Smalley, R.E., *Nature* 318, 162-63 (1985).

33. Krupke, R., Hennrich, F., Löhneysen, H. and Kappes, M.M., *Science* 301, 344 (2003).

7

Applications of Nanomaterials

7.1 Introduction

Nanotechnology offers a wide array of potential applications in diverse fields. Many possible applications and devices have been investigated. Besides, many more potential applications and new devices are being reported in literature. However, it is quite difficult to summarize all of them. The aim of this chapter is to provide some examples to illustrate the vast range of applications of nanostructures and nanomaterials.

Materials in the nano regime commonly exhibit fundamentally new behaviour. Moreover, manipulation of properties of materials at the nanoscale enables the creation of new materials and devices with enhanced or completely new characteristics and functions. Applications of nanomaterials are based on (i) the peculiar physical properties of nanosized materials, (ii) the huge surface area, and (iii) the small size that offers extra possibilities for manipulation and room for accommodating multiple functionalities.[1-7] For many applications, new materials and new properties are introduced. Forthcoming developments in nanotechnology include quantum computing, DNA computing, micro-electro-mechanical systems (MEMS), energy storage, nanobiotechnology, imaging and microscopic systems with high resolution.

It is interesting to note that the applications of nanotechnology in different fields have distinctly different demands, and thus face very different challenges, which require different approaches. For applications in nanomedicine, the major challenge is "miniaturization". For example, instruments to analyze tissues literally down to the molecular level and small machines that literally circulate within a human body to pursue pathogens and neutralize chemical toxins. Examples for potential applications of nanotechnology in energy sector are given in Table 7.1.

7.2 Molecular and Nanoelectronics

Molecular electronics is an interdisciplinary subject which focuses on using molecules in the fabrication of electronic components. This approach which reduces the size of electronic devices offers a potential method of extending Moore's law beyond the limits of conventional silicon technology. It is difficult to fabricate silicon based electronic circuitry due to the scaling limits. Molecular electronics can overcome this problem by using a single layer of molecules that self-assemble between two electrodes.[4] Thus, a self-organization of organic molecules can be used to construct electronic devices.

In molecular electronics, single molecules are expected to be able to control electron transport, which helps to explore the vast variety of molecular functions for electronic devices.[8] When the molecules are biologically active, bioelectronic devices can be developed. In molecular electronics, control over the electronic energy levels at the surface of conventional semiconductors and metals is achieved by assembling partial monolayers of molecules on the solid surfaces. Once those surfaces become interfaces, these layers exert electrostatic rather than electrodynamic control over the resulting devices, based on both electrical monopole and dipole effects of the molecules. Thus, electronic transport devices, incorporating organic molecules, can be constructed without current flow through the molecules. The simplest molecular electronics are sensors that translate unique molecular properties into electrical signals. A sensor based on field effect transistor (FET) configuration is an example. A selective membrane is inserted on the insulator surface of the FET, and this permits the diffusion of specific analyte ions and construction of a surface dipole layer at the insulator surface. Such a surface dipole changes the electric potential at the insulator surface and, thus, permits current flow through the device. Such devices are also known as ion-selective FET (ISFET) or chemical FET (CHEMFET).

Thin films attached to metal nanoparticles vary their electrical conductivity in the presence of organic vapours. This property can be used for the growth of new gas sensors. The monolayer on metal nanoparticles can reversibly adsorb and desorb the organic vapour. It will result in increase and decrease of the thickness of the monolayer, thus changing the distance between the metal cores. Since the conductivity of electrons through the monolayers depends on the distance, the adsorption of organic vapour increases the distance and leads to a sharp decrease in electrical conductivity. Schematic diagram of a nanowire-based vertical surround Gate FET is shown in figure 7.1.

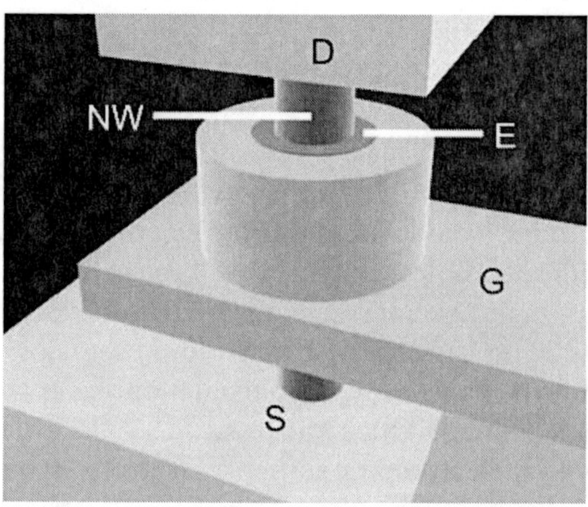

Fig. 7.1: Nanowire-based vertical surround Gate FET.[9]

The driving force for nanoelectronics is the scaling of microelectronic devices to nanoscale, which is the engine for modern information revolution. The invention of instruments in 1980s, such as STM, AFM and near-field microscopes help for the manipulation and measurement of nanostructures. The STM and AFM can be used for atom-by-atom control of materials modification, leading to atomic resolution. This potential motivated researchers to attempt the use of STM and AFM for nanolithography.

If the density of logic circuits in a chip increases to 10^8, average distance between circuits will be reduced to 1.7 μm. All of the circuit units and interconnects must be accommodated within this size limit. The required size of the devices is less than 100 nm and the width of interconnects is less than 10 nm. Quantum mechanical phenomena such as the quantization of electron energy levels, electron wave function interference, quantum tunnelling between energy levels of two nearby nanostructures and discreteness of charge carriers (e.g., single electron conductance) are dominant in this regime. The quantum devices depend on tunnelling through the forbidden energy barriers. When suitable bias voltage is applied across two nanostructures, resonance tunnelling occurs and hence tunnelling current will increase immediately.

The continuous scaling of transistors and the increase in wafer size will continue to be the trend for the semiconductor industry. The transistor gate length (feature size) has been dramatically reduced. The lateral feature size or line width of a transistor has been shrunk by several times allowing very large scale integration of transistors on a single chip. According to "International Technology Roadmap for Semiconductors", the feature sizes for lithography projected as follows: 45 nm in 2010, 32 nm in 2013 and 22 nm in 2016. Besides, the wafer size is also increased to a desired level of inches. Therefore, the performance of ICs has been dramatically improved and the cost for manufacturing has been dramatically reduced. However, the device scaling is approaching its limit. It is interesting that MOS device scaling may not be extended to below 10 nm because of physical limits such as power dissipation caused by leakage current through tunnelling. Moreover, once electronic devices approach the nano and molecular scale, the bulk properties of solids are replaced by the quantum mechanical properties of a few atoms such as energy quantization and tunnelling.

7.3 Self-Powered Nanosystems

At nanoscale, energy and technologies are greatly required for independent and continuous operations of devices such as nanobiosensors, nanorobotics and MEMS. These nanodevices need a power source to work without increasing much weight. Furthermore, it is desperate to investigate new technology to harvest energy from the environment for self-powering the nanodevices.[10-14] It is significant that nanodevices usually work at a very small power in the range of nW to μW. Therefore, the energy harvested from the environment can be enough to power the system. A human body provides several potential energy sources like mechanical energy, vibration energy, chemical energy (glucose), and hydraulic energy. If a small portion of such energy is made available as electrical energy, it is sufficient to power small biomedical devices. Figure 7.2 shows a biochemical sensor with a single SnO_2 nanobelt operating at 1.5 V and current in the range of 2-3 nA.[14] The power required to operate this nanosensor is about 5 nW. Various techniques have been developed for generating energy for mobile and wireless microelectronics using photovoltaic, thermoelectrics, mechanical vibration and piezoelectric vibration.

An overview of a few approaches that have been developed for energy harvesting at small scales is discussed in this section.

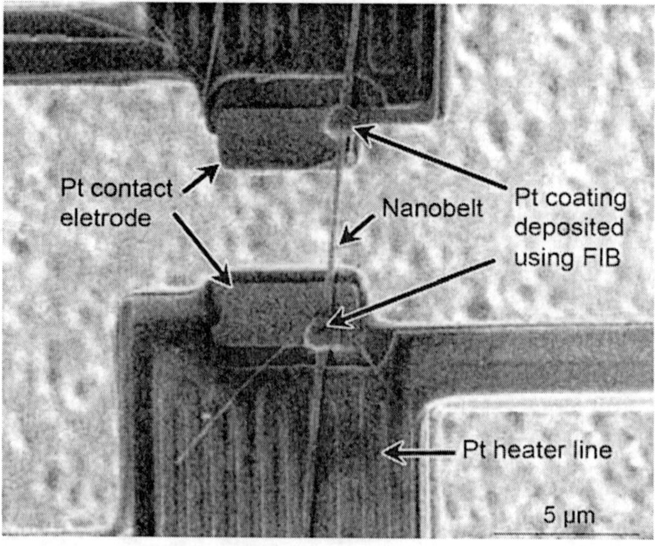

Fig. 7.2: SEM image of a biochemical sensor with a single tin oxide nanobelt.[13]

7.3.1 Microbial Fuel Cells

Micro-organisms can be used to convert bioconvertible substrates into electrical energy. In this transformation process, bacterium acts as an anode and the electrons flow from the cathode through a resistor.[15] The catalytic activities of micro-organisms can also be used to create power output from various carbohydrates and complex substrates. Direct glucose fuel cells can generate power output of about 50 μW. This power is sufficient to drive a cardiac pacemaker and its durability is approximately 150 days. Besides, glucose and oxygen present in the cells and tissues of all eukaryotic organisms are capable of generating electrical power. Thus, it is possible to explore body's own power resources, such as metabolic properties of cells, to generate sufficient energy to power clinical devices like drug delivery systems, nanorobotics and diagnostic tools.

7.3.2 Adenosine Tri-Phosphate (ATP) Energy Converter

Enzymes are considered as an alternative to powering future nanomechanical devices. Kinesin, RNA polymerase, myosin, and adenosine tri-phosphate (ATP) syntase are some of the nanoscale biological motor enzymes.[16-17] The motors are powered by ATP molecules. An ATP molecule consists of adenine, ribose and three phosphate groups, which are bonded by covalent bonds. If one phosphate group is removed, a huge amount of energy is released in the form of a reaction product called adenosine diphosphate (ADP). Similarly, the second phosphate group also releases energy in the form of adenosine monophosphate (AMP). The energy produced can be made available for chemosynthesis, locomotion and transport of ions and molecules across cell membranes. ATP is recharged with the rephosphorylation of ADP and AMP, using the chemical energy generated from the oxidation of food. Moreover, ATP can function as a rechargeable battery inside the human body.

Table 7.1: Examples for potential applications of nanotechnology in energy sector

Energy Sources	Energy Change	Energy Storage
Photovoltaic	**Gas Turbines**	**Batteries**
Nano-optimized cells (polymeric, dye, quantum dot, thin film, multiple junctions), antireflective coatings.	Heat and corrosion protection of turbine blades (ceramic or intermetallic nano-coatings) for more efficient power plants.	Optimized Li-ion batteries by nanostructured electrodes and ceramic separator foils (automobiles, mobile phones).
Wind Energy	**Electric Motors**	**Supercapacitors**
Nano composites for lighter and stronger rotor blades, wear and corrosion protection nano-coatings for bearings and power trains, etc.	Nano-composites for superconducting components in electric motors.	Nanomaterials for electrodes (CNT, metal oxides and electrolytes for higher energy density).
Geothermal	**Hydrogen Generation**	**Hydrogen**
Nano-coatings and composites for wear resistant drilling equipment.	Nano-catalysts and new processes for more efficient hydrogen generation (Photoelectrical, biophotonic, etc.).	Nanoporous materials for application in micro fuel cells for mobile electronics or in automobiles.
Biomass Energy	**Fuel Cells**	**Fuel Tanks**
Yield optimization by nano-based precision farming. Nanosensors will control the release and storage of pesticides and nutrients.	Nano-optimized membranes and electrodes for efficient fuel cells for applications in automobiles or mobile electronics.	Gas tight fuel tanks based on nano-composites for reduction of hydrocarbon emissions.
Nuclear	**Thermoelectrics**	**Adsorptive Storage**
Nano-composites for radiation shielding and protection, long-term option for nuclear reactors.	Nanostructured compounds for efficient thermoelectric power generation.	Nanoporous materials for reversible heat storage in buildings and heating nets.

7.3.3 Thermoelectric Generator

Thermoelectric generator is based on the Seebeck Effect, which states that an electric potential is developed between two dissimilar metals that form a junction and are at different temperatures. The potential difference is proportional to the temperature difference between the two ends. In nanotechnology, thermoelectric power is one of the most significant types of energies, which are used to power nanodevices.[18] Properties such as thermal conductivity and electrical conductivity are high in one dimensional nanomaterials (e.g., Bi and BiTe) and are very helpful to promote thermal power.

7.3.4 Piezoelectric Nanogenerator

Investigators discovered new techniques for converting mechanical energy (body movement, muscle stretching), vibration energy (acoustic/ultrasonic wave), and hydraulic energy (body fluid and blood flow) into electric energy, which can be used to power nanodevices.[19-20] A study based on aligned ZnO nanowires (NWs) grown on a conductive solid substrate is shown in figure 7.3(a). Here, AFM with Si tip coated with Pt film is used for the measurements. In contact mode, a constant normal force (say ~ 5 nN) is kept between the tip and sample surface [figure 7.3(b)]. The tip is scanned over the top of ZnO nanowire. For each contact position, voltage output image contains many sharp output peaks, as shown in figure 7.3(c).

Both piezoelectric and semiconducting properties are joined together for generating piezoelectric charges in the NW. For a vertical ZnO NW [figure 7.3(a)], the deflection of the NW by AFM tip produces a strain field. The outer surface becomes tensile and inner surface compressive [figure 7.3(b)]. The piezoelectric effect may create a piezoelectric field Ez along the NW. This field is parallel to the NW at the outer surface and anti-parallel to the NW at the inner surface [figure 7.3(c)]. As a result, across the width of the NW at the top end, the electric potential distribution from the compressed to the stretched side surface is approximately between –Vs to +Vs.

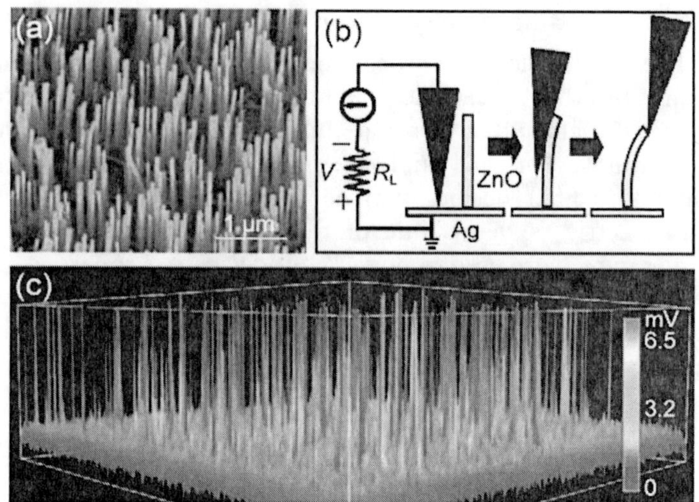

Fig. 7.3: (a) SEM images of aligned ZnO nanowires grown on sapphire substrate. (b) Schematic diagram of a piezoelectric nanogenerator. (c) AFM output voltage images of the NW arrays.[13]

7.3.5 Fibre-based Nanogenerator

It is worth to note that the energy that is available from biological systems can be captured by developing suitable fibre based nanogenerators. These nanogenerators are made on flexible and foldable substrates and target at low frequency excitations (~ 10 Hz). The design of the nanogenerator is based on the mechanism of utilizing zig zag electrode or using an array of metal wires.[21] Nanogenerator that is made from a collection of metal nanowires is shown in figure 7.4(a). The metal nanowires act like an array of AFM tips. If the metal nanowire arrays brush across the ZnO nanowire arrays, the metal nanowires induce mechanical deformation to the ZnO nanowires and collect the charges. ZnO nanowire arrays can be grown at 80°C on substrates of any fibre and the metal nanowire arrays are built by metal coating of ZnO nanowire arrays grown on Kevlar fibre [figure 7.4(b)]. Entangling two fibres, with and without Au coating, sets the principle of the fibre-based nanogenerator as shown in figure 7.4(c). A repeated sliding between the two fibres produces output current due to the deflection and bending of the ZnO nanowires. This fibre based nanogenerator has the potential for harvesting energy from body movement, muscle stretching, light wind, and vibration.

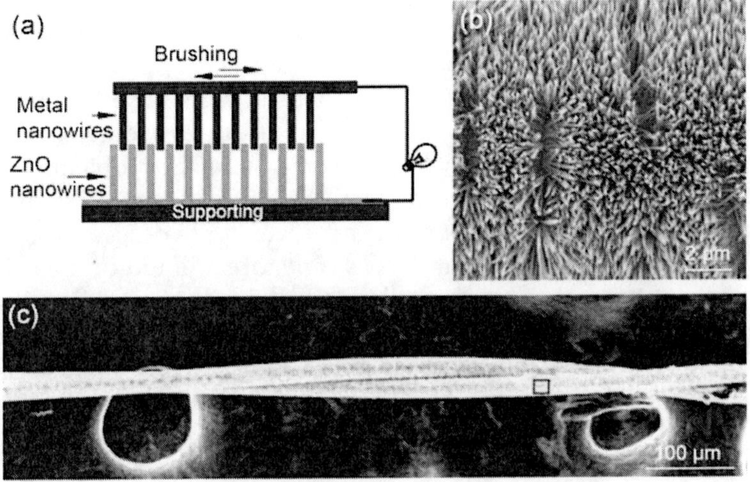

Fig. 7.4: (a) Schematic model of the nanogenerator, (b) Aligned ZnO nanowire arrays grown on surfaces of a fibre, (c) SEM image of a pair of entangled fibres.[13]

7.3.6 Solar Photovoltaic Cells

Photovoltaic cells convert solar energy into electricity. They use photon-electron excitation process to release electrons in semiconductors. It may not release any pollutants to the environment. Solar cells have a great future in providing power for rural communities, homes and businesses. But, the technology still remains expensive and is also not very efficient. The typical efficiency of a solar cell is only up to 15-20 per cent, however, nanotechnology has the potential to make solar cells more efficient. In a conventional solar cell, when solar photons strike the cell, they release electrons in the semiconductor to cause an electric current. Even though solar photons carry enough energy to release several electrons, only one electron per solar photon is released. As a result, solar cells operate at very low efficiency. It is worth to note that use of quantum dots as semiconductor allows solar energy to release multiple electrons. Research reports shows that the use of quantum dots for solar cells can enhance the efficiency of solar cells to about 40-45 per cent.

7.4 Hydrogen Storage

Hydrogen is found abundant in nature and is an ideal fuel source. The advantage of hydrogen based energy source is that its combustion product is water vapour and and also hydrogen can be easily regenerated. Besides, this fuel is clean, efficient and safe.[22-23] The storing of hydrogen gas is the main problem with hydrogen fuel. Carbon nanostructures like CNTs and graphite nanofibres (GNFs) are promising candidates for hydrogen storage. It is worth mentioning that CNTs can store a liquid or a gas in the inner cores through a capillary effect due to their cylindrical and hollow geometry and nanoscale diameters. In 1997, Dillon and his co-workers investigated that SWNTs have a high degree of hydrogen storage capacity. They found that hydrogen can be condensed to high capacity (5~10 wt %) inside the narrow pores of SWNTs. The adsorption of H_2 in SWNT has been examined with temperature programmed desorption (TPD) spectroscopy and has been observed that physical adsorption of hydrogen occurred within the hollow space of SWNTs.

Chambers and his co-workers reported that hydrogen can be stored in graphite nanofibres. Their result shows that various forms of GNFs, such as tubular, platelet, and herringbone have the capacity of 11, 45, and 67 wt % H_2, respectively at room temperature and a pressure of about 12 MPa. They also found that GNFs have special structure, which produces a material composed of nanopores that hold hydrogen molecules. The pore walls can expand to accommodate hydrogen molecules.

7.5 Nanomedicine

A promising and fast growing field for the application of nanotechnology is in the practice of medicine that is often referred to as nanomedicine. Nanotechnology has found numerous applications in medical field.[24-25] The strategies for generating multifunctional nanoparticles share common approaches, whether the nanoparticles are nanoshells, carbon nanotubes, dendrimers, iron oxides, quantum dots or other nanoparticles. In addition to these platform nanoparticles, there are a large variety of nanoparticles constructed of other types of materials. They all involve encapsulation, covalent conjugation, or noncovalent adsorption of various moieties to allow the nanoparticles to recognize and locate the tumour, deliver a load or kill the tumour cells, and permit visualization and imaging. Some important applications of nanotechnology in the field of medicine are discussed in this section.

7.5.1 Drug Delivery System

The prime objective of the drug delivery system is to deliver sufficient dosage of drug to the infected part of the body. But, most of these systems do not succeed to work effectively. As the size of the drug particles is large, they may not be absorbed by cells. Sometimes these drug particles are insoluble or they may cause tissue damages. However, nanoparticles are easily absorbed by the cells due to their extremely small size. Furthermore, they are fully soluble and may not cause tissue damages. It is interesting that the efficiency of the drug delivery system can be increased a number of times by integrating nanoparticles with them. For example, most of the present day cancer treatment methods cause large scale damage of normal cells. The reason is that, while eradicating the infected cells they also destroy the normal cells. However, nanotechnology based drug delivery systems can effectively treat cancer without damaging the nearby cells and tissues. Since the nanoparticles are smaller than the body cells, they can easily carry the drug to the infected cells of the body. CNT based drug delivery systems and their working are discussed in section 6.4.2.3.

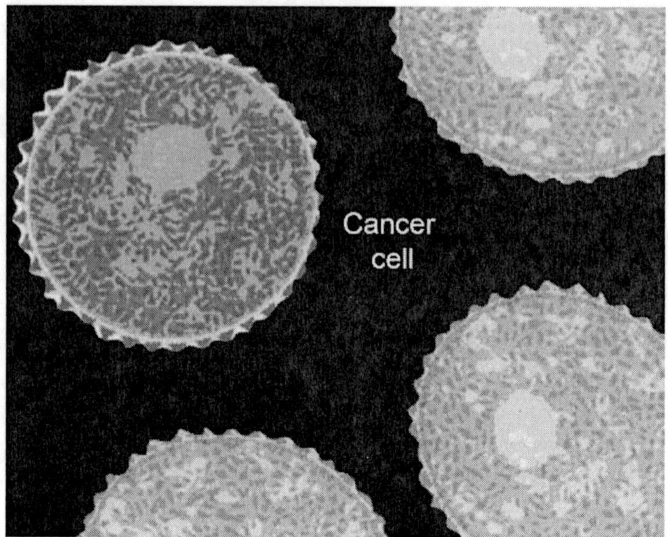

Fig. 7.5: Nanoparticles used for molecular imaging of malignant lesions.

Biopharmaceuticals are used in the treatment of cancer and other life-threatening diseases. They are peptides or protein molecules that trigger multiple reactions in human body. Their efficiency can be improved by bonding them with nanoparticles. The nanoparticles can carry the biopharmaceuticals directly to the tumour site without any damage to normal cells and tissues.

Nanotube, nanoshell and mesoporous nanoparticle have hollow and porous structures, and facile surface functionalization. These properties make them ideal for drug delivery vehicles. Their inner void can load a large amount of drug and the open ends of pores act as gates, which can manage the drug release. Moreover, unlike spherical nanoparticles, hollow structures isolate drug payload from the environment and can transport the payloads safely into the cell. SWNTs show optical absorbance in the near infrared (NIR) range. Making use of this property, one can use SWNT to trigger drug/gene release by illumination of NIR light. The absorbance of NIR light can generate heat that triggers the release of drugs or genes from the nanotube surface. MWNTs can be used as nanoneedles because they pass through the cell membrane without any cell damage.

Dendrimers are spherical polymers; their diameter is usually less than 5 nm. They have polymer branches with large surface area to accommodate therapeutic agents and targeting molecules. A typical dendrimer starts with an ammonia (NH_3) core that is reacted with acrylic acid to produce a tri-acid molecule. It is also possible to use sugar or other molecules as the starting core, so long as they have multiple, identical reaction sites. The resultant molecule is then reacted with ethylenediamine to produce a tri-amine, and is called generation 0 (G0) product. The tri-amine reacts with acrylic acid to produce a hexa-acid, and then reacts with ethylenediamine to produce a hexa-amine (G1), and so on. This alternation of reaction with acrylic acid then with ethylenediamine continues until the desired generation is reached. Thus, it is possible to create a surface consisting of multiple amines or multiple acids, and these two kinds of surfaces provide the means of attaching different functional components.

7.5.2 Nanorobots

The creation of nanoscale devices for improved therapy and diagnostics is a very significant application of nanomedicine. Such nanoscale devices are known as nanorobots or simply as nanobots. The discovery of nanorobots can revolutionize the world of medicine. The size of nanorobots is very small and hence, they can be simply entered into the blood vessels, organs, tissues and cells of the body. These miniature devices are able to detect diseases and infections of the body, and will also be able to repair internal injuries and wounds. They can be programmed to carry out some specific tasks. Nanobots applied to medicine would be able to make out a cancer cell or an invading virus, and do some task to fix the target. For example, nanorobots release drug locally or bind to the target and protect it from virus activity. Further in future, gene replacement, tissue regeneration or nanosurgery are all possibilities as the technology becomes more mature and sophisticated. Schematic diagram of a blood swimming nanorobot is shown in figure 7.6.

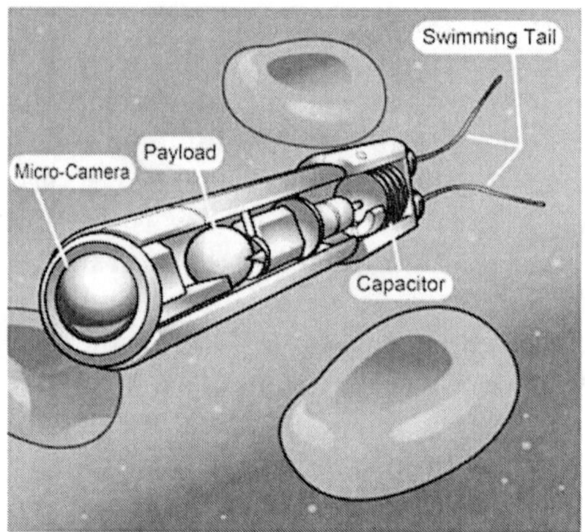

Fig. 7.6: Model of a blood swimming robot.

Source: www.electronics.howstuffworks.com

Cancer primarily occurs due to mutation and hence, there is a change in the genetic information stored in the DNA. The affected cells divide repeatedly and cause the formation of tumours. Problems come to the fore only when the condition becomes unmanageable. Nanobots can enter into the infected cell and will repair the damaged DNA. Besides, they can also make out cells that are growing abnormally and thus, the detection of cancer at an early stage is possible.

7.5.3 Cellular Imaging

Quantum dots are nanocrystals, their diameter ranges from 2 to 10 nm. They are made of semiconductors, the most common being cadmium selenide capped by zinc sulfide (CdSe/ZnS). Quantum dots usually consist of 10–50 atoms, and they confine electron-hole pairs to a discrete quantized energy level. When ultraviolet light is incident on QDs, they will be excited and fluoresce with different colours depending on their size, which determines the energy level of the quantum dot. Smaller particles emit light in the blue region, while larger particles emit in the red end of the visible region. QDs are currently being used as probes for high resolution molecular imaging of cellular components and for tracking a cell's activities and movements inside the body.

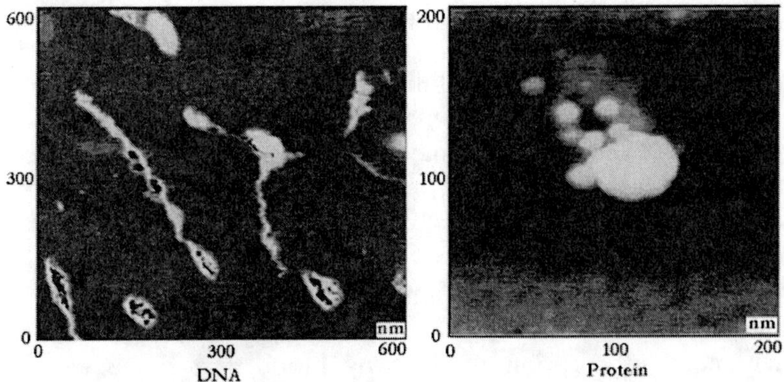

Fig. 7.7: Images of DNA and protein.

Source: www.ipt.arc.nasa.gov

Cellular labelling using quantum dots (QDs) has made progress, attracted the greatest interest, and also reached certain stage of commercialization. Recently, numerous reports have appeared describing the use of biofunctionalized QDs that have different sizes to label cells and provide multicolour imaging.

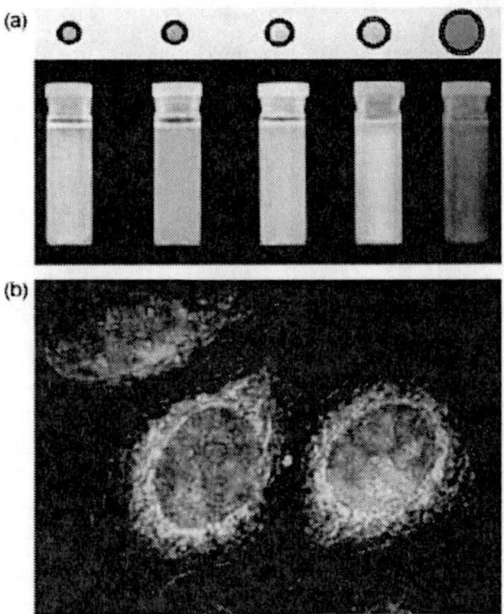

Fig.7.8: (a) The fluorescent emission from red to blue by QDs with different sizes excited by a UV lamp, (b) The multicolour imaging of cells using QDs.

Source: www.blog.aperio.com

Such applications highlighted the advantages offered by QDs as a unique type of fluorophores. For example, the antibody conjugated QDs show capability of multicolour labelling of live cells. Cellular multicolour imaging using the QDs is shown in figure 7.8. Moreover, QDs have been used for *in vivo* biological imaging. Researchers have demonstrated the cancer targeting and imaging by using CdSe@ZnS core-shell QDs that give red emission.

For *in vivo* biological imaging applications of QDs, the fluorescent emission wavelength should be within the region where the absorption of light by blood and tissue will be minimum, but is still detectable by the instruments. Thus, the QDs emit around 700-900 nm in the near-infrared (NIR). Furthermore, the NIR QDs can permit imaging deeper than conventional near-infrared dyes. The development of NIR QDs is progressing rapidly. Several cadmium based materials with NIR emission, such as CdTe@CdSe core-shell Type II QDs hold great promise in several applications. However, the non-cadmium, NIR QDs may be better for biomedical applications owing to their enhanced biocompatibility.

Nanoshells consist of a silica core coated with a thin gold shell, as discussed in section 2.4. It is possible to manipulate the thickness of the core and the outer shell, in order to absorb and scatter specific wavelengths of light in the visible and NIR region. This property has led the potential application in thermal ablation therapy. For example, gold nanoshells absorb NIR light (~800 nm) and can produce intense heat. The property to scatter light has the potential for cancer imaging.

7.5.4 In Vivo Communication

Natural triggering and induced triggering are the two types of communication inside the body. Neurons are the carriers for the information transfer in natural triggering (e.g., antibody generation). But in induced triggering (e.g., targeted drug delivery), nanomaterials are the information carriers. The communication medium will be an aqueous system inside the body. Nanodevices for *in vivo* communication should be designed with minimum friction.

7.5.5 Superparamagnetic Nanoparticles and Cancer Therapy

The most interesting applications of superparamagnetic nanoparticles are related to medicine and biology. Superparamagnetic nanoparticles refer to iron oxide particles or magnetite (Fe_3O_4) particles that are less than 10 nm in diameter. It is possible to activate magnetic nanoparticles remotely by electromagnetic fields. Magnetic nanoparticles can also be used to thermal treatment of cancer cells. When an alternating field is applied to superparamagnetic nanoparticles, they undergo Brownian relaxation, in which heat is generated by the rotation of particles in the field. However, 0.01 to 0.1 per cent iron oxide concentrations are required to raise the tissue to critical temperatures for thermal ablation. These concentrations are hard to achieve via intravenous administration. Nowadays, superparamagnetic nanoparticles are used in clinical thermotherapy of locally recurrent prostate cancer. Thermotherapy can render cancer cells more susceptible to the effects of radiation and cause some apoptosis. Superparamagnetic nanoparticles can also help to enhance the contrast in nuclear magnetic resonance (NMR) imaging.

7.6 Biological Applications

One important branch of nanotechnology is nanobiotechnology. Nanobiotechnology includes (i) the use of nanostructures as highly sophisticated scopes, machines or materials in biology or medicine, and (ii) the use of biological molecules to assemble nanoscale structures. There are numerous biological applications of nanotechnology, but only a few is discussed in this section.

Molecular recognition is one of the most interesting capabilities of many biological molecules.[26-28] Some biological molecules can recognize and bind to other molecules with extremely high selectivity and specificity. For molecular recognition applications, antibodies and oligonucleotides are widely used as receptors. Antibodies are protein molecules created by the immune systems of higher organisms that can recognize a virus as a hostile intruder or antigen, and bind to it such that the virus can be destroyed by other parts of the immune system. Oligonucleotides, known as single stranded deoxyribonucleic acid (DNA), are linear chains of nucleotides, each of which is composed of a sugar backbone and a base. There are four types of bases: adenine (A), cytosine (C), guanine (G), and thymine (T). The molecular recognition ability of oligonucleotides arises from two characteristics. One is that each oligonucleotide is characterized by the sequence of its bases, and another is that base A only binds to T and base C only to G. This makes the binding of oligonucleotides highly selective and specific.

Antibodies and oligonucleotides are typically attached to the surface of nanocrystals through (i) thiol-gold bonds to gold nanoparticles, (ii) covalent linkage to silanized nanocrystals with bifunctional crosslinker molecules, and (iii) a biotin-avidin linkage, where avidin is adsorbed on the particle surface. When a nanocrystal is attached or conjugated to a receptor molecule, it is "tagged". Nanocrystals conjugated with a receptor can now be "directed" to bind to positions where ligand molecules are present, which "fit" the molecular recognition of the receptor. This facilitates a set of applications including molecular labelling. For example, when gold nanoparticles aggregate, a colour change will occur from ruby-red to blue. This property enables the development of really sensitive colorimetric methods of DNA analysis.

7.6.1 Biosensors

Nanotechnology is playing a vital role in the development of biosensors.[27-29] A wide variety of nanomaterials with different properties have found very useful applications in many kinds of biosensors for biomedical fields. As nanomaterials in the range 1-100 nm exhibit unique physical, chemical and electronic properties, make them suitable for designing novel sensing devices. Also these dimensions are similar to many cellular objects, including DNA, cell surface receptors and viruses. Many types of nanomaterials of different sizes and compositions can play various roles in different sensing systems. A biosensor is an integrated receptor-transducer device, which can provide selective quantitative or semi-quantitative analytical information.

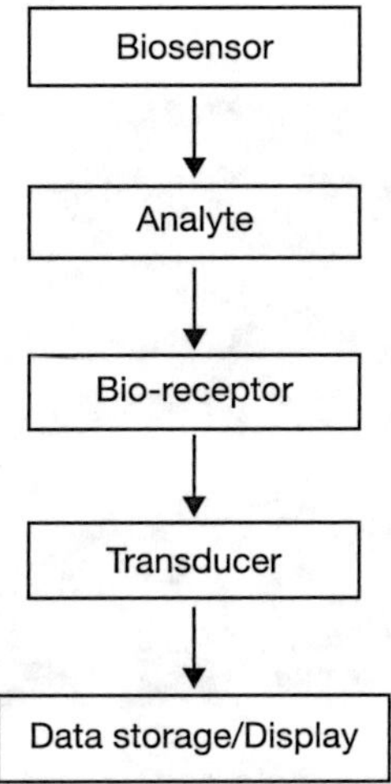

Fig. 7.9: Schematic flow chart showing the main components of a biosensor.

Basically, a biosensor consists of a biological recognition element (bioreceptor), which acts upon a biochemical mechanism and a transducer relying on electrochemical, mass, optical or thermal principles. Typical bioreceptors are enzymes, antibodies, micro-organisms and nucleic acids. A number of signal transduction methods are employed in biosensors. These methods include optical, radioactive, electrochemical, piezoelectric, magnetic, micromechanical and mass spectrometric technique. The characteristic trait of a biosensor is the direct spatial contact between the bioreceptor and transducing element. A flow chart showing the conceptual working of a biosensor is shown in figure 7.9. The analytical devices composed of a bioreceptor can directly be interfaced to a signal transducer which together relates the concentration of an analyte (or group of related analytes) to a measurable response. The bio-reaction converts the substrate to product. The transducer measures the physical change that occurs with reaction at the bioreceptor and then converts it to an electrical signal. This analog signal is converted into a digital signal and then processed using a microprocessor stage. Finally, the processed data is fed to a display or storage device. Figure 7.10 shows biosensors, such as nanoscale cantilever for detecting biomarkers of cancer.

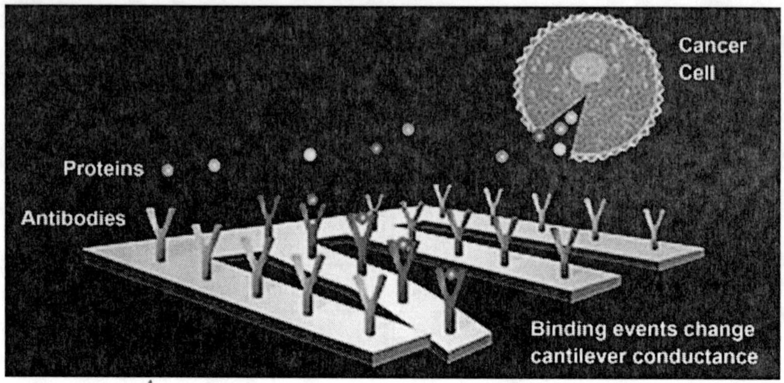

Fig. 7.10: Nanoscale cantilevers for detecting biomarkers of cancer.

Source: www.nano.cancer.gov

The basic functions of biosensing nanomaterials can be mainly classified as:

(i) Immobilization of biomolecules: Nanoparticles can adsorb biomolecules strongly, and help the immobilization of biomolecules in biosensor construction due to the large surface area and high surface free energy.

(ii) Catalysis of electrochemical reactions: Many nanomaterials, especially metal nanoparticles show excellent catalytic properties. When these nanoparticles are introduced into biosensors, they can reduce overpotentials of many analytically important electrochemical reactions.

(iii) Enhancement of electron transfer: The electrical conduction of nanoparticles can be used for improving the electron transfer between the active centres of enzymes and electrodes, acting as electron transfer mediators or electrical wires.

(iv) Labelling biomolecules: The labelling of biomolecules, such as antigen, antibody and DNA with nanoparticles plays an increasingly important role in developing sensitive electrochemical biosensors. Biomolecules labelled with nanoparticles can retain their bioactivity and interact with their counterparts.

(v) Acting as reactant: Based on the active chemical properties of nanoparticles and using these nanoparticles as special reactants, some novel electrochemical analysis systems can be constructed.

7.7 Catalysis

Catalysis process modifies the rate of a chemical reaction due to the participation of a substance called a catalyst. Catalyst can be used to change one type of molecule to another at lower temperatures. It makes the reaction more effective. Catalysis is one of the significant applications of nanoparticles. Nanoparticles have large surface area to volume ratio, unique surface structure and electronic states. These properties help them to stimulate and promote more catalytic activity. Nanocatalyst find applications in fuel cells, production of chemicals, catalytic converters and photocatalytic devices.

7.8 Pollution Control and Filtration

Nanocatalysts and nanostructured membranes are highly useful in reducing air and water pollutions. As discussed in section 7.7, nanocatalysts can be used to enable a chemical reaction more effectively. They can be used to convert vapours escaping from vehicles or industrial plants into harmless gases. The larger surface area of the nanoparticles helps them to interact with more chemicals simultaneously, which enhances the catalystic activity. Nanostructured membranes separate greenhouse gases like methane or carbon dioxide from industrial plants exhaust streams. They can trap moving gases much faster than those by conventional gas separation techniques. For example, carbon nanotube membranes have the potential to separate high volumes of greenhouse gases from other gases. Methane gas from underground coal mining can be separated before extracting the coal, using CNT membrane.

Nanofiltration is an effective method for the removal of ions or metal contaminants. Magnetic nanoparticles have the potential to eliminate heavy metal contaminants from waste water by using magnetic separation techniques. The efficiency of the nanoparticles to absorb the contaminants is very high and is comparatively inexpensive compared to usual filtration techniques. Nanotechnology based water treatment devices are already available on the market.

7.9 Photonic Nanocrystals and Integrated Circuits

Photonic nanocrystals are synthetic materials with a patterned periodic dielectric or metallic structures. They have photonic band-gap similar to electronic band gap in semiconductors. Photons with energies that lie within the band gap cannot be propagated, while a defect such as point defect, line defect or planar defect results an allowed propagation state within the band gap. These defects are like optical cavities, waveguides, or mirrors, and provide a new mechanism for light control and advancement of all optical integrated circuits. Photonic integrated circuits could pack individual components heavily. The tighter confinement and unique dispersion properties can contribute new applications in nonlinear (optical) devices and low power communication devices.

7.10 Nanomaterials in Communication Sector

Nanotechnology offers a broad range of promising applications in telecommunications, computing and networking sectors. An overview of various types of communications and their recent advancements are discussed in this section.

7.10.1 Nanolasers

The complex interaction between nanometre structures and light can introduce new technology for devices and sensors. Nanoscience researchers are investigating light emission from a semiconductor nanowire. It has a diameter of about 10-100 nanometres and a few micrometres long, which functions as a laser. Lasers constructed from arrays of nanowires have found potential applications in communications and sensing.

7.10.2 Electronic Communication and Informatics

Communication via electronic means is termed as electronic communication. Internet, fax, satellite, television, computers and networks are examples of electronic communication. Nanotechnology enabled systems offer faster and powerful information handling techniques and devices. Nanotechnology interfaced with electronics would result in single-molecule and single-electron based transistors. High performance devices can be constructed from these kinds of transistors. Informatics deals with the structure, creation, management, storage, retrieval, dissemination and transfer of information. Informatics mainly consists of a processor which translates one programme to another and is the backbone of the communication system. Each processor contains a definite number of transistors with specific functions linked with it. The present day communication system is based on silicon technology and the technology is entering into a near molecular regime, as the size has come down to 25 nm. If suitable technology can be explored to manipulate light at small scales, photonic technologies will take over silicon technology. Remarkable electronic properties of CNTs are also found appropriate for developing an alternate technology.

7.10.3 Quantum Computers and Quantum Structure Electronic Devices

The principles of quantum mechanics have become more and more imperative at the nano regime, as discussed in chapter 3. The concept of quantum computing was introduced in 1970s. It is fully based on quantum physics that allows the atoms and nuclei to work together as quantum bits or qubits and to be the processor and memory. Qubits can execute calculations exponentially and quicker than conventional systems. A quantum computer makes information processing by using quantum mechanical phenomena, such as superposition and entanglement to perform operations on data. Unlimited processing power and secure data communication are the two important features of quantum computing.

In quantum computation, atoms are the basis for computation. In this system, atoms can have three states (0 or 1 or 01), where the last one is a superposition of the first two states. But logic system only exist in two states 0 or 1. A pair of qubits can be in any quantum superposition of 4 states, and three qubits in any superposition of 8 states. In general, a quantum computer with 'n' qubits can be in an arbitrary superposition of 2^n different states simultaneously. But, an ordinary computer can only be in one of these 2^n states at a time. Quantum computers can be operated by manipulating those qubits with a fixed sequence of quantum logic gates and the sequence of logic gates is called a quantum algorithm.

One of the prime objectives of nanotechnology is to construct three-dimensionally confined quantum structure electronic devices (QSDs) such as quantum wire and quantum dot devices for better performance. QSDs can confine electrons to a very narrow regime, < 20 nm. Quantum well lasers for telecommunications, high electron mobility transistors (HEMTs) for low noise and high gain microwave applications, and vertical cavity surface emitting lasers (VCSELs) for data communications are some of the successful QSDs.

7.10.4 Optical Communication

Optical communication system uses light as the transmission medium. An optical communication system consists of a transmitter, a channel and a receiver. The transmitter encodes a message into an optical signal, the channel carries the signal to its destination and the receiver recovers the original signal from the optical signal. Usually, optical fibre is used as the channel for optical communications. The transmitters in optical fibre communication system are commonly light-emitting diodes (LEDs) or laser diodes, which converts electrical signal into light signal. Optical fibres transmit infrared light with less attenuation and dispersion. So IR light is preferred over visible light. Data rates for lasers are much higher as compared to low data rates of LEDs. It has been reported that under suitable voltages, thin films composed of ferroelectric materials form "nanobubbles". They have the potential to store lots of information in a tiny space. This could force technologies from computers and portable electronics to radio frequency identification devices.

7.10.5 Satellite Communication

An artficial satellite is a radio relay station in orbit above the earth. It receives signal from earth stations, amplifies it and then redirects back to earth. Semiconductor quantum dots with a small amount of substrate materials are ideal for satellite communications, because they cover the entire wavelength region from ultraviolet to far infrared. Besides, quantum dots have other advantages like low power consumption, high modulation range for high speed applications and better thermal stability. In GaAs quantum dot lasers are already used in communication satellites. Ken Teo and his co-workers, University of Cambridge,

> There are typically 50 microwave amplifiers built on a satellite. Each amplifier weighs about 1 kg and its linear dimension is about 30 cm. A minimum of ten lakhs rupees is required to send one kilogram payload into space. If the weight and size of the microwave devices are reduced with CNTs, it is possible to send extra payload even at low cost.

have proposed to use an array of CNTs to create a device, which replaces conventional microwave amplifiers. CNT, the new electron source, promises to revolutionize satellite communications.

7.10.6 Communication Devices

7.10.6.1 Micro Electro Mechanical Systems

A technology that combines computers with tiny mechanical devices is called micro-electromechanical systems (MEMS). Microelectronic ICs are the brains of the system. Basically, a MEMS device is composed of microcircuitry on a tiny semiconductor IC chip into which some tiny mechanical devices such as sensors, gears, valves or actuators have been embedded.[1] For example, flight wings can be fabricated with sensors which can sense and react to the air flow. This is possible by changing the wing surface resistance by creating indefinite number of tiny wing flaps. MEMS devices are embedded systems and do not work alone. For example, MEMS help a component to perform higher level functions, such as controlling the fuel to air mixture in a car's engine. Both MEMS and nanotechnology terms deal with microminiaturized objects, but, they are different. MEMS deals with creating devices at micrometre dimensions, whereas nanotechnology deals with manipulating atoms at the nanometre scale. MEMS have the ability to gather data, process it, find suitable course of action, and then act as a trigger by communicating through an electronic interface. These capabilities allow them to work with smart devices, such as collision avoidance systems and wireless handsets.

MEMS combine various science disciplines, including physics, bioinformatics, biochemistry, electrical engineering, optics and electronics. MEMS devices offer amazing applications in variety of fields. They are used to create pressure, temperature, chemical and vibration sensors, light reflectors and switches. In automotive industry, they are incorporated into airbag and vehicle control. In medical field, they are used to control medication dosing and medical devices such as pacemakers. This technology is also used to make ink jet print heads, micro actuators for read/write heads. MEMS mirrors reflect the input signal to an appropriate output port of an optical switch as shown in figure 7.11. This method is expected to be effective for building photonic switches which are essential parts of the communication devices.

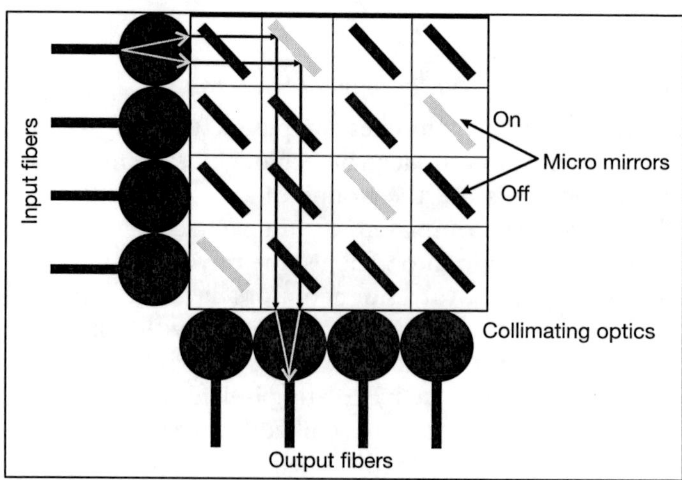

Fig. 7.11: Schematic model of MEMS based optical switch.

7.10.6.2 Smart Dust

Smart dust is an emerging technology, which is composed of hundreds to thousands tiny wireless sensors or motes.[30] As the name implies, it is smart enough to communicate with other sensors and is very small to fit even on the head of a pin. Each motes has to sense and monitor environmental conditions and to communicate with other devices. Each mote consists of a tiny computer with a power supply, sensors, analog circuitry, bi-directional communication system and a programmable microprocessor. Advances in various scientific areas, such as miniaturization, integration, energy management, optical communications and MEMS led to the manufacturing of small sensors and other smart dust components. Figure 7.12 shows the schematic diagram of a smart dust mote.

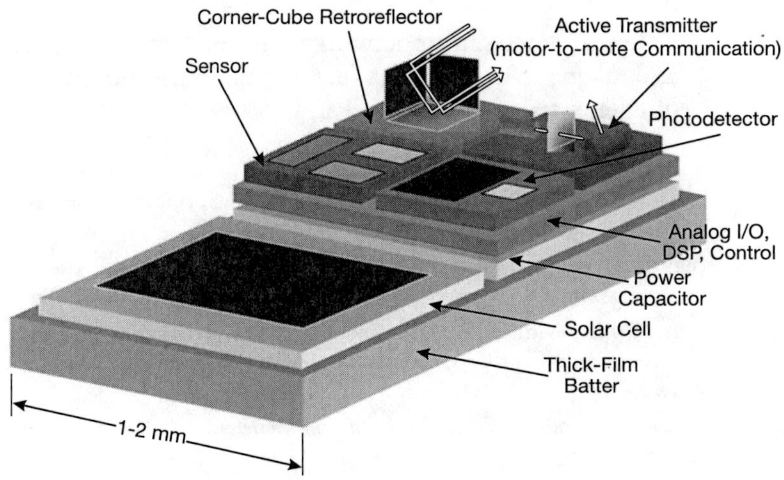

Fig. 7.12: Smart dust mote with sensor, optical receiver, optical transmitters, signal-processing and control circuitry, and power sources.

A smart dust can act as a distributed autonomous sensor network. The motes gather necessary information from the neighbouring environment and submit these report via a communication network. Dust motes can have direct communication with a base station transceiver, or peer-to-peer communication can be made between dust motes. Smart dusts have numerous applications in industry and military. Some useful applications for smart dust are listed below:

- Collecting data for meteorological, geophysical, or planetary research.
- Tracking the movements of birds, small animals, and even insects.
- Virtual keyboard (Acceleration sensing glove)—2 axis accelerometer on top of the each finger on a data glove.
- Smart office spaces—temperature, humidity, and environmental comfort sensors.
- Monitoring product quality using temperature, humidity, pressure sensors.
- Defense related sensor networks using acoustic, vibration and magnetic field sensors (e.g., tracking the movements of enemy troops).

REFERENCES

1. Bhusion, B., *Handbook of nanotechnology* (NY: Springer-Heidelberg, 2004).
2. Knauth, P. and Schoonman, J., *Nanostructured Materials: Selected Synthesis Methods, Properties and Applications* (Springer, 2002).
3. Rao, C.N.R., Müller, A. and Cheetham, A.K., *The Chemistry of Nanomaterials: Synthesis, Properties and Applications* (John Wiley & Sons, 2004).
4. Nalwa, H.S. (Ed.), *Handbook of Nanostructured Materials and Nanotechnology* (Academic Press: San Diego, 1999).
5. Brechignac, C., Houdy, P. and Lahmani, M. (ed.), *Nanomaterials and Nanochemistry* (Springer: NY, 2007).
6. Edelstein, A.S. and Cammarata, R.C. (ed.), *Nanomaterials: Synthesis, Properties and Applications* (IOP Publishing: Bristol and Philadelphia, 1996).
7. Guozhong, Cao, *Nanostructures and Nanomaterials: Synthesis, Properties, and Applications* (Imperial College Press: London, 2004).
8. Mahler, G., May, V. and Schreiber, M. (eds.), *Molecular Electronics: Properties Dynamics and Applications* (NY: *Marcel Dekker*, 1996).

9. Meyyappan, M., *Nanotechnology: Opportunities and Challenges, NASA Ames Research Center* (www.ipt.arc.nasa.gov).
10. Special issue on Sustainability and Energy, *Science*, 2007, Feb. 9.
11. Special issue on Harnessing Materials for Energy, *MUS Bull.*, 2008, 33 (4).
12. Wang, Z.L., Self-powering nanotech. *Scientific American* 2008, 82, 87.
13. Wang, Z.L., Energy Harvesting for Self-powered nanosystems, *Nano Res*, 1:1-8 (2008).
14. Yu, C., Hao, Q., Saha, S., Shi, L., Kong, X. and Wang, Z.L., Integration of metal oxide nanobelts with Microsystems for nerve agent detection. *Appl. Phys. Lett.* 2005, 86, 63101.
15. Bond, D.R., Holmes, D.E., Tender, L.M. and Lovley, D.R., Electrode-reducing microorganisms that harvest energy from marine sediments. *Science*, 2002, 295, 483-85.
16. Yasuda, R., Noji, H., Kinosita, K. Jr. and Yosida, M., F1-ATPase is a highly efficient molecular motor that rotates with discrete 120° steps, *Cell*, 1998, 93, 1117-24.
17. Noji, H., Yasuda, R., Yoshida, M. and Kinosita, K., Jr. Direct observation of the rotation of F1-ATPase, *Nature*, 1997, 386, 299-302.
18. Shi, L., Li, D., Yu, C., Jang, W., Kim, D., Yao, Z., Kim, P. and Majumdar, A., *J. Heat Transfer*, 2003, 125, 881-88.
19. Wang, Z.L. and Song, J.H. Piezoelectric nanogenerators based on zinc oxide nanowire arrays. *Science*, 2006, 312, 242-46.
20. Zhou, J., Xu, N.S. and Wang, Z.L., *Adv. Mater.*, 2006, 18, 2432-35.
21. Qin, Y., Wang, X.D. and Wang, Z.L., Microfiber-nanowire hybrid structure for energy scavenging. *Nature*, 2008, 451.
22. Michael, J. (ed.), *Carbon Nanotubes: Properties and Applications* (Taylor & Francis, 2006).
23. Yury, Gogtsi (ed.), *Carbon Nanomaterials* (Taylor and Francis, 2006).
24. Gohel, M.C. and Parikh, R.K., *Targeted Drug Delivery Systems* 7, 3 (2009).
25. Bergveld, P., IEEE Trans. Biomed. *Eng.* BME19, S342 (1972).
26. Kourosh, Kalantar-zadeh and Benjamin, Fry, *Nanotechnology Enabled Sensors* (Springer, 2008).
27. Jha, A.R., *Nanotechnology-Based Sensors and Devices for Communication, Medical and Aerospace Applications*, CRC Press, (Taylor and Francis, 2008).

28. Thomas, Varghese, Sukirtha, T.H. and Joseph, Sunny, *Biosensors Based on Nanomaterials, Nanotechnology in Biology and Medicine*. Singh, S.K. (Ed.), (New Delhi: Studium Press (India), 2011.
29. Pradeep, T., *Nano: The Essentials* (Tata McGraw Hill: New Delhi, 2007).
30. Song, Y., *Optical Communication Systems for Smart Dust* (Thesis), Virginia Polytechnic Institute and State University, 2002.

Glossary

Aerogel: A silicon-based foam composed mostly of air.

Aerosol: A suspension of fine particles (0.01-10 microns) of a solid or liquid in a gas.

Adenosine Triphosphate (ATP): A chemical compound that functions as fuel for biomolecular nanotechnology having the formula, $C_{10}H_{16}N_5O_{13}P_3$.

Angstrom Unit: Unit of length used to measure atoms and molecules, 10^{-10} meter or 0.1 nanometer.

Assembler: A general-purpose device for molecular manufacturing capable of guiding chemical reactions by positioning molecules.

Atomic Layer Deposition (ALD): A self-limiting, sequential surface chemistry that deposits conformal thin-films of materials onto substrates of varying compositions.

Atomic Force Microscope (AFM): A scanning probe microscopy instrument capable of revealing the structure of samples.

Atomic Manipulation: Manipulating atoms with the tip of an STM.

Atomistic Simulations: Computations of the properties of nanosystems using computer simulation techniques.

Bio-assemblies or Biomolecular Assemblies: Assemblies containing several biologically significant molecules such as protein units, DNA loops, lipids and various ligands.

Biopolymer: A polymer found in nature (examples, DNA and RNA).

Biovorous: An organism capable of converting biological material into energy for sustenance.

Biomedical Nanotechnology: See nanomedicine.

BioMEMS: MEMS used in medicine.

BioNEMS: Biofunctionalized nanoelectromechanical systems.

Biomimetics: The study of the structure and function of biological substances to make artificial products that mimic the natural ones.

Biomimetic Chemistry: Chemistry of new molecules, molecular assemblies, and macromolecules having biomimetic functions.

Biomimetic Materials: Materials that imitate, copy, or learn from nature.

Biopolymeroptoelectromechanical Systems [BioPOEMS]: Combining optics and microelectromechanical systems for biological applications.

Biosensor: A sensor used to detect a biological substance (for example: bacteria, blood gases, or hormones).

Bottom Up: Building larger objects from smaller building blocks like atoms and molecules.

Buckminsterfullerene: See Fullerenes. A broad term covering the variety of buckyballs and carbon nanotubes that exist. Named after the architect Buckminster Fuller, who is famous for the geodesic dome.

Buckyballs: A molecules made up of 60 carbon atoms arranged in a series of interlocking hexagonal shapes, forming a structure similar to a soccer ball (see fullerenes).

Carbon: A nonmetallic element found in all living things. Carbon is part of all organic compounds and, in combined form of many inorganic substances. Diamonds, graphite, and fullerenes are pure forms of carbon.

Carbon Nanotubes (CNTs): Long, thin cylinders of carbon having remarkable physical properties.

Cell Pharmacology: Delivery of drugs by medical nanomachines to exact locations in the body.

Cell Repair Machine: Molecular and nanoscale machines with sensors, nanocomputers and tools, programmed to detect and repair damage to cells and tissues.

Characterization: Analysis of critical features of an object or concept.

Chemical Vapour Deposition (CVD): A technique used to deposit coatings, where chemicals are first vapourized, and then applied using an inert carrier gas such as nitrogen.

Chemisorption: The process by which a liquid or gas is chemically bonded to the surface of a solid.

Cognotechnology: Convergence of nanotech, biotech and IT, for remote brain sensing and mind control.

Colloids: Very fine solid particles that will not settle out of a solution or medium.

Composite: A material made from two or more components that has properties different from the constituent materials. Composite materials have two phases: matrix (continuous) phase, and dispersed phase (particulates, fibres).

Computational Nanotechnology: The modelling and simulation of complex nanometer-scale structures.

Computronium: A highly efficient matrix for computation, such as dense lattices of nanocomputers or quantum dot cellular automata.

Dendrimers: A polymer with tiny molecular structure that interacts with cells, enabling scientists to probe, diagnose, cure diseases or manipulate cells at nanoscale.

Diamondoid: Strong stiff structures containing dense, three dimensional networks of covalent bonds, formed chiefly from first and second row atoms with a valence of three or more.

Dip Pen Nanolithography: An AFM-based soft-lithography technique.

DNA (deoxyribonucleic acid): The molecule that encodes genetic information, found in the cell's nucleus.

DNA Chip: Gene Chip and DNA Microchip. A purpose built microchip used to identify mutations or alterations in a gene's DNA.

Dopeyballs: Superconducting Buckyballs. They have the highest critical temperature of any known organic compound.

Drug Delivery: The use of physical, chemical and biological components to deliver controlled dose of medicine to a diseased cell.

Ecosystem Protector: A nanomachine for mechanically removing selected imported species from an ecosystem to protect native species.

Electron Microscopy: The visual examination of very small structures with a device that forms greatly magnified images of objects by using electrons rather than light to create an image.

Encapsulation: The condition of being enclosed or the process of enclosing.

Entanglement: From quantum mechanics, entanglement is a relationship between two objects in which they both exhibit superposition but once the state of one object is measured, the state of the other is also known.

Entropy: A measure of the disorder of a closed system. The second law of thermodynamics states that the entropy (and disorder) increases as time moves forward.

Epitaxy: The growth of a crystal layer of one mineral on the crystal base of another mineral in such a manner that the crystalline orientation of the layer mimics that of the substrate.

Field Effect: The local change from the normal value produced by an electric field in the charge-carrier concentration of a semiconductor.

Field Emission: The emission of electrons from the surface of a metallic conductor into a vacuum (or into an insulator) under influence of a strong electric field.

Fluorescence Spectroscopy: A technique to measure the interaction of radiant energy with matter by passing emitted fluorescent light through a monochromator to record the fluorescence emission spectrum.

Fractal: A mathematical construct that has a fractional dimension.

Fractal Mechatronic Universal Assembler: A machine that is capable of assembling any chemical from a generic descriptions of the properties required of the chemical.

Fullerenes: Fullerenes are a molecular form of pure carbon discovered in 1985. They are cage-like structures of carbon atoms, the most abundant form produced is buckminsterfullerene (C_{60}), with 60 carbon atoms arranged in a spherical structure.

Genegeneering: Genetic engineering.

Giant Magnetoresistance (GMR): It results from subtle electron-spin effects in ultra-thin 'multilayers' of magnetic materials, which cause huge changes in their electrical resistance when a magnetic field is applied. GMR is 200 times stronger than ordinary magnetoresistance.

GNR Technologies: Convergence of Genetic Engineering, Nanotechnology, and Robotics.

Heisenberg Uncertainty Principle: A quantum-mechanical principle with the consequence that the position and momentum of an object cannot be precisely determined. The Heisenberg principle helps determine the size of electron clouds, and hence the size of atoms.

Immune Machines: Medical nanomachines designed for internal use, especially in the bloodstream and digestive tract, able to identify and disable intruders such as bacteria and viruses.

***In vitro*:** An experiment done in an artificial environment.

***In vivo*:** A medical experiment done within a living object.

Infrared (IR) Spectroscopy: A technique in which infrared light is passed through matter and some of the light is absorbed by inciting molecular vibration. The difference between the incident and the emitted radiation reveals structural and functional data about the molecule.

Khaki Goo: Military nanotechnology.

Knowbots: Knowledge robots.

Langmuir-Blodgett: The name of a nanofabrication technique used to create ultrathin films (monolayers and isolated molecular layers), the end result of which is called a "Langmuir-Blodgett film".

LCD (Liquid Crystal Display): The predominant technology used in flat panel displays. Alignment of liquid crystals can be altered with electric current. By using green, blue and red emitting crystals, a full colour display can be made.

LEDs (Light Emitting Diodes): The recombination of the electron hole pairs in certain semiconductors emits light of a particular frequency (hence a particular colour) depending on the physical characteristics of the semiconductor used.

Lithography: The process of imprinting patterns on materials.

Limited Assembler: Assembler capable of making only certain products; faster, more efficient, and less liable to abuse than a general-purpose assembler.

Low-dimension Structures: Quantum wells, quantum wire and quantum dots.

Mechanochemistry: The direct, mechanical control of molecular structure formation and manipulation to form atomically precise products.

MEMS—Microelectromechanical Systems: The term used to describe micron scale electrical/mechanical devices.

Mesoscale: A device or structure larger than the nanoscale (10^{-9} m) and smaller than the megascale.

Molecular Assembler: Also known as an assembler, a molecular assembler is a molecular machine that can build a molecular structure from its component building blocks.

Molecular Beam Epitaxy (MBE): Process used to make compound (multi-layer) semiconductors. Consists of depositing alternating layers of materials, layer by layer, one type after another.

MOCVD (Metal-Organic Chemical Vapour Deposition): A technique for growing thin layers of compound semiconductors in which metal-organic compounds are decomposed near the surface of a heated substrate wafer.

Molecular Integrated Microsystems (MIMS): Microsystems in which functions found in biological and nanoscale systems are combined with manufacturable materials.

Molecular Electronics (ME): [Moletronics] Any system with atomically precise electronic devices of nanometer dimensions, especially if made of discrete molecular parts rather than the continuous materials found in today's semiconductor devices.

Molecular Manufacturing: Manufacturing using molecular machinery, giving molecule-by-molecule control of products and by-products via positional chemical synthesis.

Molecular Medicine: Studying molecules as they relate to health and disease, and manipulating those molecules to improve the diagnosis, prevention, and treatment of disease.

Molecular Nanotechnology (MNT): Thorough, inexpensive control of the structure of matter based on molecule-by-molecule control of products and byproducts; the products and processes of molecular manufacturing, including molecular machinery.

Molecular Systems Engineering: Design, analysis, and construction of systems of molecular parts working together to carry out a useful purpose.

Molecular Wire: A molecular wire—the simplest electronic component—is a quasi-one-dimensional molecule that can transport charge carriers (electrons or holes) between its ends.

MOLMAC: Molecular machine.

Monomer: The units from which a polymer is constructed.

Moore's Law — Coined in 1965 by Gordon Moore, states that the number transistors packed into an integrated circuit will double in every two years.

MSTM: Magnetic STM.

Nanarchist: Someone who circumvents government control to use nanotechnology, or someone who advocates this.

Nanarchy: The use of automatic law-enforcement by nanomachines or robots, without any human control.

Nanite: Machines with atomic-scale components.

Nanobeads: Polymer beads with diameters of between 0.1 to 10 micrometers. Also, called nanodots, nanocrystals and quantum beads.

Nanobot: Nanorobot (See Nanite).

Nanobubbles: Tiny air bubbles on colloid surfaces.

Nanocomposites: Materials that result from the intimate mixture of two or more nanophase materials.

Nanocomputer: A computer made from components (mechanical, electronic, or otherwise) built at the nanometer scale.

Nanochondria: Nanomachines existing inside living cells, participating in their biochemistry (like mitochondria) and/or assembling various structures.

Nanocones: Nonplanar graphitic structures. Carbon-based structures with fivefold symmetry that form due to disclination defects in two-dimensional graphene sheets.

NEMS—Nanoelectromechanical Systems: A generic term to describe nanoscale electrical/mechanical devices.

Nanoelectronics: Electronics on a nanometer scale, whether made by current techniques or nanotechnology; includes both molecular electronics and nanoscale devices.

Nanofacture: The fabrication of goods using nanotechnology.

Nanofilters: Filters for the separation of molecules, such as proteins or DNA, for research in genomics.

Nanofluidics: Controlling nanoscale amounts of fluids.

Nanolithography: Writing nanoscale patterns. *See* Lithography.

Nanomachine: An artificial molecular machine of the sort made by molecular manufacturing.

Nanomaterials: Subdivided into nanoparticles, nanofilms and nanocomposites. The focus of nanomaterials is a bottom up approach to structures and functional effects whereby the building blocks of materials are designed and assembled in controlled ways.

Nanomedicine: The area of research focusing on the development of a nanoscale technologies for disease diagnosis, treatment, and prevention.

Nanometer (nm): 10^{-9} m.

Nano-Optics: Interaction of light and matter on the nanoscale.

Nanoparticles: Nanoscale spherical or capsule shaped structures.

Nanopharmaceuticals: Nanoscale particles used to modulate drug transport for drug uptake and delivery applications.

Nanoprobe: Nanoscale machines used to diagnose, image, report on, and treat disease within the body.

Nanoreplicators: A set of nanomachines capable of exponential replication.

Nanorods or Carbon Nanorods: Formed from multi-wall carbon nanotubes, a nanoscale material with unique and promising physical properties.

Nanoropes: Nanotubes connected and strung together.

Nanoscale: 1-100 nanometer range.

Nanoscopic Scale: Same as nanoscale.

Nanosensors: Nanoscale sensors.

Nanoshells: Nanoscale metal spheres, which can absorb or scatter light at virtually any wavelength.

Nanostructures: Structures made from nanomaterials.

Nanotechnology: A manufacturing technology able to inexpensively fabricate most structures consistent with natural law, and to do so with molecular precision.

Nanotubes: Long, thin cylinders of carbon, discovered in 1991 by S. Iijima. These large macromolecules are unique for their size, shape, and remarkable physical properties. They can be thought of as a sheet of graphite (a hexagonal lattice of carbon) rolled into a cylinder. The physical properties are still being discovered.

OLED or Organic LED: LEDs made from carbon-based molecules.

Quantum: A small discrete package of light energy.

Quantum Computer: A computer that takes advantage of quantum mechanical properties such as superposition and entanglement resulting from nanoscale, molecular, atomic and subatomic components.

Quantum Confined Atoms (QCA): Atoms caged inside nanocrystals. May find uses in clear-glass sunglasses, bio-sensors, and optical computing.

Quantum Dots: Nanoscale crystalline structure made from cadmium selenide that absorbs white light and then re-emits a specific colour.

Quantum Mirage: A nanoscale property that may allow information to be transferred through use of the wave property of electrons. Thus, quantum computers might not require wires as we know them.

Quantum Tunnelling: When electrons pass through a barrier, without overcoming it or breaking it down.

Quantum Wire: Another form of quantum dot, but unlike the single-dimension "dot", a quantum wire is confined only in two dimensions—that is it has "length", and allows the electrons to propagate in a "particle-like" fashion.

Qubit: The quantum computing analog to a bit. Qubits exhibit superposition. Thus, unlike normal bits, qubits can be both 1 and 0 at the same time.

Raman Spectroscopy: Analysis of the intensity of Raman scattering, in which light is scattered as it passes through a material medium and suffers a change in frequency and a random alteration in phase. The resulting information is useful for determining molecular structure.

Scaffold: Three-dimensional biodegradable polymers engineered for cell growth.

Scanning Electron Microscopy (SEM): An electron microscope that images a sample by scanning it with a high-energy beam of electrons in a raster scan pattern.

Scanning Force Microscope (SFM): An instrument able to image surfaces to molecular accuracy by mechanically probing their surface contours.

Scanning Near Field Optical Microscopy: A method for observing local optical properties of a surface that can be smaller than the wavelength of the light used.

SPM: Scanning Probe Microscope, including AFM and STM, in which effect of interaction of a sharp probe with the sample is found to infer atomic structure of the material.

Scanning Tunnelling Microscope (STM): An instrument used to image conducting surfaces to atomic accuracy has been used to pin molecules to a surface.

Self-assembly: In chemical solutions, self-assembly (also called Brownian assembly) results from the random motion of molecules and the affinity of their binding sites for one another.

Single-walled Carbon Nanotubes (SWNT): See Nanotubes and buckyballs.

Smart Dust: Tiny, micro-machines fitted with wireless communication devices—that measure light and temperature among other things, such as environmental monitoring, health, security, distributed processing and tracking.

Smart Materials: Materials and products capable of relatively complex behaviour due to the incorporation of nanocomputers and nanomachines.

Sol-gel Process: A chemical synthesis technique for preparing gels, glasses, and ceramic powders generally involving the use of metal alkoxides.

Superconductor: An object or substance that conducts electricity with zero resistance.

Technocyte: A nanoscale artificial device (especially a nanite) in the human bloodstream used for repairs, cancer protection, as an artificial immune system or for other uses.

Thin Film: A film one molecule thick; often called monolayer.

Top Down Moulding: Carving and fabricating small materials and components by using larger objects such as our hands, tools and lasers, respectively.

Transmission Electron Microscopy (TEM): The use of electron high energy beams to achieve magnification close to atomic observation.

UV/ViS (Ultraviolet-Visible) Spectroscopy/Spectrophotometry: Method to determine concentrations of an absorbing species in solution. This technique uses light in the visible and adjacent near ultraviolet (UV) and near infrared (NIR) ranges to achieve this quantitative analysis.

van de Waals Force: Weak intermolecular attractive forces.

Wet Nanotechnology: Bionanotechnology, where nanosystems function in the presence of water.

Index

L

M

N